Prepared in collaboration with the University of Wisconsin–Milwaukee

Development and Application of a Groundwater/Surface-Water Flow Model using MODFLOW-NWT for the Upper Fox River Basin, Southeastern Wisconsin

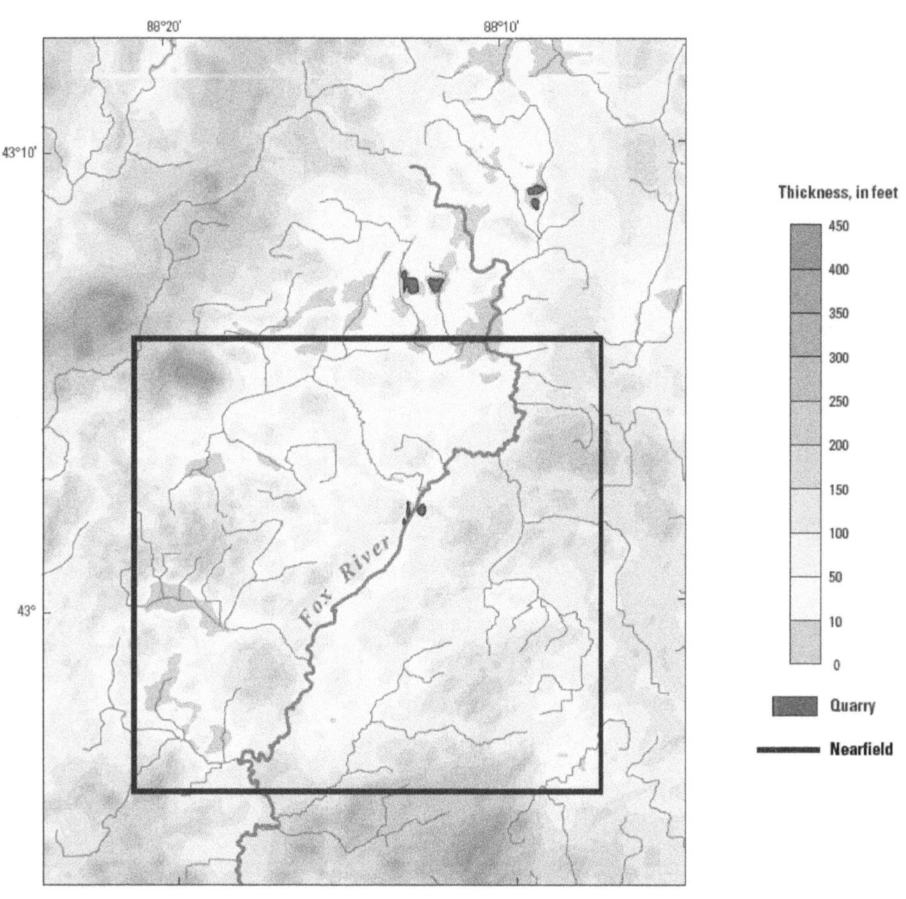

Scientific Investigations Report 2012–5108

U.S. Department of the Interior
U.S. Geological Survey

Cover. Map showing thickness of shallow aquifer system in model domain—unconsolidated deposits (see figure 4a in report for details).

Development and Application of a Groundwater/Surface-Water Flow Model using MODFLOW-NWT for the Upper Fox River Basin, Southeastern Wisconsin

By D.T. Feinstein, M.N. Fienen, J.L. Kennedy, C.A. Buchwald, and M.M. Greenwood

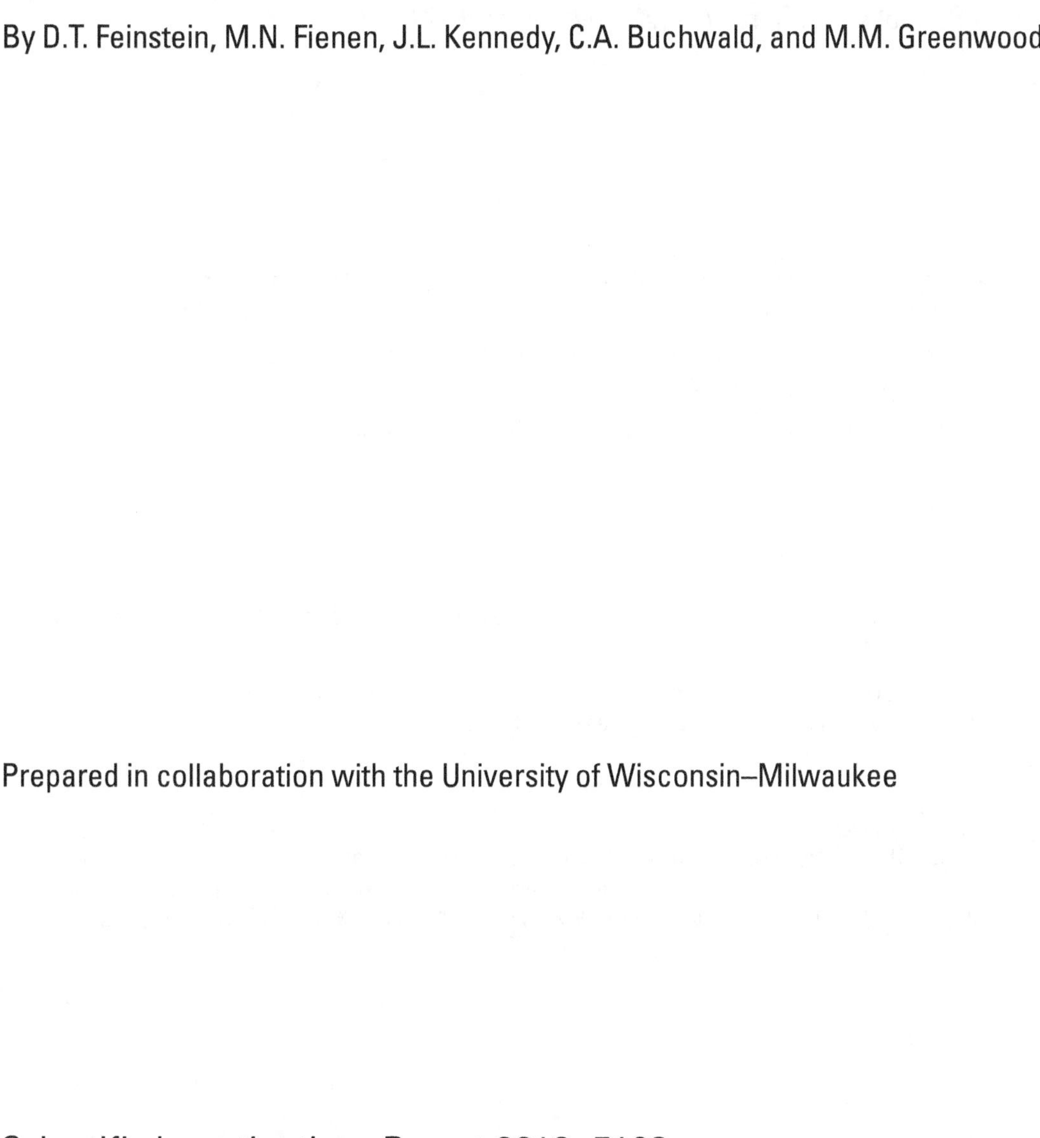

Prepared in collaboration with the University of Wisconsin–Milwaukee

Scientific Investigations Report 2012–5108

U.S. Department of the Interior
U.S. Geological Survey

U.S. Department of the Interior
KEN SALAZAR, Secretary

U.S. Geological Survey
Marcia K. McNutt, Director

U.S. Geological Survey, Reston, Virginia: 2012

For more information on the USGS—the Federal source for science about the Earth, its natural and living resources, natural hazards, and the environment, visit http://www.usgs.gov or call 1–888–ASK–USGS.

For an overview of USGS information products, including maps, imagery, and publications, visit http://www.usgs.gov/pubprod

To order this and other USGS information products, visit http://store.usgs.gov

Suggested citation:
Feinstein, D.T., Fienen, M.N., Kennedy, J.L., Buchwald, C.A., and Greenwood, M.M., 2012, Development and application of a groundwater/surface-water flow model using MODFLOW-NWT for the Upper Fox River Basin, southeastern Wisconsin: U.S. Geological Survey Scientific Investigations Report 2012–5108, 124 p.

Acknowledgments

The development and application of the Upper Fox model depended on contributions from several groups of scientists.

At the U.S. Geological Survey, Richard Niswonger provided indispensable assistance in facilitating the use of the recently released MODFLOW-NWT code. John Walker furnished base-flow estimates and statistical support. David M. Ely and David Pollock reviewed the report and it benefited greatly from their remarks. Bobbie Louthian provided a thorough editorial review supplemented by insightful comments from John Eggleston.

The funding for the project originated in a grant from the Brico Fund (Milwaukee, Wisconsin). Peter McAvoy, now adjunct at the University of Wisconsin–Milwaukee (UWM) School of Freshwater Science, helped coordinate the research and shape the research strategy. At UWM, Professors Douglas Cherkauer (emeritus) and Timothy Grundl were closely involved with every step of model development and application – their participation was essential. As part of her graduate work for Prof. Cherkauer, Bonnie Bills tirelessly compiled and interpreted well logs and surface-water data. The project also benefited from geophysical investigations conducted by Prof. William Kean (emeritus) and his graduate student, Michael Baierlipp.

Several members of the Wisconsin groundwater community provided vital information for the project:

- John Jansen (Cardno ENTRIX) and Ted Powell (Ruekert and Mielke, Inc.) located and forwarded pumping test and other records; John also offered good suggestions on modeling approaches;
- Michael Hahn (Southeastern Wisconsin Regional Planning Commission) and representatives of the Waukesha Water Utility provided important hydrogeologic and water-use data;
- Researchers at the Wisconsin Geological and Natural History Survey (Kenneth Bradbury, Madeline Gotkowitz, and David Hart) were, as always, helpful in talking through ideas.

Contents

Figures

Tables

Conversion Factors

Inch/Pound to SI

Multiply	By	To obtain
	Length	
foot (ft)	0.3048	meter (m)
mile (mi)	1.609	kilometer (km)
	Area	
square foot (ft^2)	0.09290	square meter (m^2)
square mile (mi^2)	259.0	hectare (ha)
square mile (mi^2)	2.590	square kilometer (km^2)
	Volume	
gallon (gal)	3.785	liter (L)
gallon (gal)	0.003785	cubic meter (m^3)
million gallons (Mgal)	3,785	cubic meter (m^3)
cubic foot (ft^3)	0.02832	cubic meter (m^3)
	Flow rate	
cubic foot per second (ft^3/s)	0.02832	cubic meter per second (m^3/s)
cubic foot per day (ft^3/d)	0.02832	cubic meter per day (m^3/d)
gallon per minute (gal/min)	0.06309	liter per second (L/s)
million gallons per day (Mgal/d)	0.04381	cubic meter per second (m^3/s)
inch per year (in/yr)	25.4	millimeter per year (mm/yr)
	Hydraulic conductivity	
foot per day (ft/d)	0.3048	meter per day (m/d)
	Hydraulic gradient	
foot per mile (ft/mi)	0.1894	meter per kilometer (m/km)

Temperature in degrees Celsius (°C) may be converted to degrees Fahrenheit (°F) as follows:

°F=(1.8×°C)+32

Temperature in degrees Fahrenheit (°F) may be converted to degrees Celsius (°C) as follows:

°C=(°F-32)/1.8

Vertical coordinate information is referenced to the North American Vertical Datum of 1929 (NAVD 29).

Elevation, as used in this report, refers to distance above the respective vertical datum.

Horizontal spatial reference for the model grid is in Wisconsin Transverse Mercator projection,

North American Datum of 1983 (NAD 83).The grid coordinates are in units of feet.

*Transmissivity: The standard unit for transmissivity is cubic foot per day per square foot times foot of aquifer thickness [(ft^3/d)/ft^2]ft. In this report, the mathematically reduced form, foot squared per day (ft^2/d), is used for convenience.

Development and Application of a Groundwater/Surface-Water Flow Model using MODFLOW-NWT for the Upper Fox River Basin, Southeastern Wisconsin

By D.T. Feinstein, M.N. Fienen, J.L. Kennedy, C.A. Buchwald, and M.M. Greenwood

Abstract

The Fox River is a 199-mile-long tributary to the Illinois River within the Mississippi River Basin in the states of Wisconsin and Illinois. For the purposes of this study the Upper Fox River Basin is defined as the topographic basin that extends from the upstream boundary of the Fox River Basin to a large wetland complex in south-central Waukesha County called the Vernon Marsh. The objectives for the study are to (1) develop a baseline study of groundwater conditions and groundwater/surface-water interactions in the shallow aquifer system of the Upper Fox River Basin, (2) develop a tool for evaluating possible alternative water-supply options for communities in Waukesha County, and (3) contribute to the methodology of groundwater-flow modeling by applying the recently published U.S. Geological Survey MODFLOW-NWT computer code, (a Newton formulation of MODFLOW-2005 intended for solving difficulties involving drying and rewetting nonlinearities of the unconfined groundwater-flow equation) to overcome computational problems connected with fine-scaled simulation of shallow aquifer systems by means of thin model layers.

To simulate groundwater conditions, a MODFLOW grid is constructed with thin layers and small cell dimensions (125 feet per side). This nonlinear unconfined problem incorporates the streamflow/lake (SFR/LAK) packages to represent groundwater/surface-water interactions, which yields an unstable solution sensitive to initial conditions when solved using the Picard-based preconditioned-gradient (PCG2) solver. A particular problem is the presence of many isolated wet water-table cells over dry cells, causing the simulated water table to assume unrealistically high values. Attempts to work around the problem by converting to confined conditions or converting active to inactive cells introduce unacceptable bias. Application of MODFLOW-NWT overcomes numerical problem by smoothing the transition from wet to dry cells and keeps all cells active. The simulation is insensitive to initial conditions and the water-table trend is smooth across layers. The MODFLOW-NWT code permits rigorous calibration and also robust application of the model to transient scenarios. Runtimes on a 64-bit computer are kept reasonably short by use of updated initial conditions and informed choices of solver parameters.

The shallow aquifer system consists of unconsolidated material of varying thickness over Silurian dolomite. The unconsolidated material, largely of glacial origin, contains fine-textured and coarse-textured deposits that vary in permeability over short distances. This study at least partly encompasses the inevitable uncertainty in the hydraulic conductivity zones by developing two models—one favors the continuity of fine-grained deposits and a second favors the continuity of coarse-grained deposits. The separate calibration processes for the fine-favored and coarse-favored models using MODFLOW-NWT and the nonlinear regression algorithms in the parameter estimation (PEST) code produce distinct parameter values for hydraulic conductivity zones, storage parameters, and streambed conductance zones.

Both models are applied to a hypothetical scenario involving 27 "riparian" wells completed adjacent to the river channel and open to the shallow aquifer systems along a 10-mile stretch of the Fox River. The results suggest that a riparian well system withdrawing about 9 million gallons per day would induce about one-third to one-half its total discharge from the river, and that this riverbank inducement would appreciably limit drawdown around the hypothetical wells.

1. Introduction

The Fox River Basin (fig. 1*A*) spreads over multiple counties. Several of these counties, such as McHenry and Kane Counties in Illinois and Waukesha County in Wisconsin, are undergoing rapid urbanization and consequent stresses on their water-supply systems (CH2MHill, 2002; Groschen and others, 2004; Kay and others, 2006; Meyer and others, 2009). Water-quality considerations that affect groundwater supplies also in some instances complicate the outlook for a sustainable water supply. The city of Waukesha, for example, in the face of radium exceedances in wells tapping the deep Cambrian-Ordovician aquifer system, is evaluating alternative sources of water such as increased withdrawals from shallow wells and a possible diversions from Lake Michigan (Southeastern Wisconsin Regional Planning Commission, 2010). Other communities in the Fox River Basin are likely to face similar choices in efforts to augment their water-supplies in coming years. The possibility that shallow pumping will be increased throughout the basin could imply decreased base flow to the Fox River and its tributaries and increased stresses on lakes and wetlands, especially in the context of uncertain climate trends.

Resource-management issues involving water supply and environmental impact can be evaluated with quantitative tools such as groundwater-flow models. This report describes the development and example application of a groundwater-flow model for the uppermost part of the Fox River Basin in Waukesha County. The model focuses on groundwater/surface-water interactions in the shallow part of the groundwater system, including the effect of shallow wells on water-table elevations and flows through the surface-water network.

The objectives for the study are: to develop a baseline study of groundwater conditions and groundwater/surface-water interactions in the shallow aquifer system of the Upper Fox River Basin as first step in a possible regional study of the Fox River Basin in the States of Wisconsin and Illinois; to develop a tool for evaluating possible alternative water-supply options for communities in Waukesha County in the context of the need to substitute or augment deep aquifer water tainted by radium exceedances; and to contribute to the methodology of groundwater-flow modeling by applying the recently published MODFLOW-NWT code (Niswonger and others, 2011) in an effort to overcome computational problems connected with fine-scaled simulation of shallow aquifer systems by means of thin model layers. Groundwater-flow models that focus on shallow conditions commonly are compromised by numerical instabilities prompted by dewatering of model cells. This study implements a new methodology aimed at overcoming these limitations.

1.1 Purpose and Scope

This report documents the construction and calibration of groundwater-flow models used to evaluate groundwater-flow patterns in the shallow aquifer system within the Upper Fox River Basin in southeastern Wisconsin. The report

- contributes to the understanding of the hydrogeology of the study area by developing two flow models that offer distinct interpretations of the unconsolidated (mostly glacial) material—one favors the continuity of fine-grained sediments and the second favors the continuity of coarse-grained deposits;
- integrates hydrogeologic and hydrologic data with the hydrogeology to define sources and sinks of groundwater, including recharge, boundary fluxes, interactions with surface water, and discharge to wells and quarries;
- quantifies groundwater/surface-water interactions at a scale fine enough to study the effect of pumping on flow to and from individual surface-water features such as the Fox River and its tributaries; and
- demonstrates the use of the MODFLOW-NWT code (Niswonger and others, 2011) to overcome solver limitations and, thereby, improve the capacity of flow models to simulate shallow groundwater conditions and groundwater/surface-water interactions.

The Upper Fox River Basin models are designed to provide a platform for studying how development pressures could be managed to minimize effects on local recharge and discharge conditions in the Upper Fox River Basin. The models could anticipate the effects of climate change on base flows; water levels in the basin could be mitigated. The model could aid in understanding how groundwater-supply systems in the basin take advantage of riverbank inducement to minimize not only drawdown but also base-flow reduction.

1.2 Description of Upper Fox River Basin Study Area

The Fox River is a 199-mile (mi)-long tributary to the Illinois River within the Mississippi River Basin in the states of Wisconsin and Illinois (fig. 1*A*). The headwaters of the Fox River rise at the northern boundary of Waukesha County in southeastern Wisconsin. The river flows generally southward defining a topographic basin that includes parts of Waukesha, Racine, Kenosha, and Walworth Counties in Wisconsin and continues southward into northeastern Illinois. For the purposes of this study the *Upper Fox River Basin* is defined as the topographic basin that extends from the upstream boundary of the Upper Fox River Basin to a large wetland complex in south-central Waukesha County called the Vernon Marsh (fig. 1*B*). The north/south extent of the Upper Fox River Basin is approximately 20 mi and the basin area is approximately 207 square miles (mi^2). It is bordered by the Lower Rock River Basin to the northwest and west, by the Milwaukee River Basin to the east, by the Pike-Root Rivers Basin to the southeast, and by the continuation of the Fox River Basin to the south.

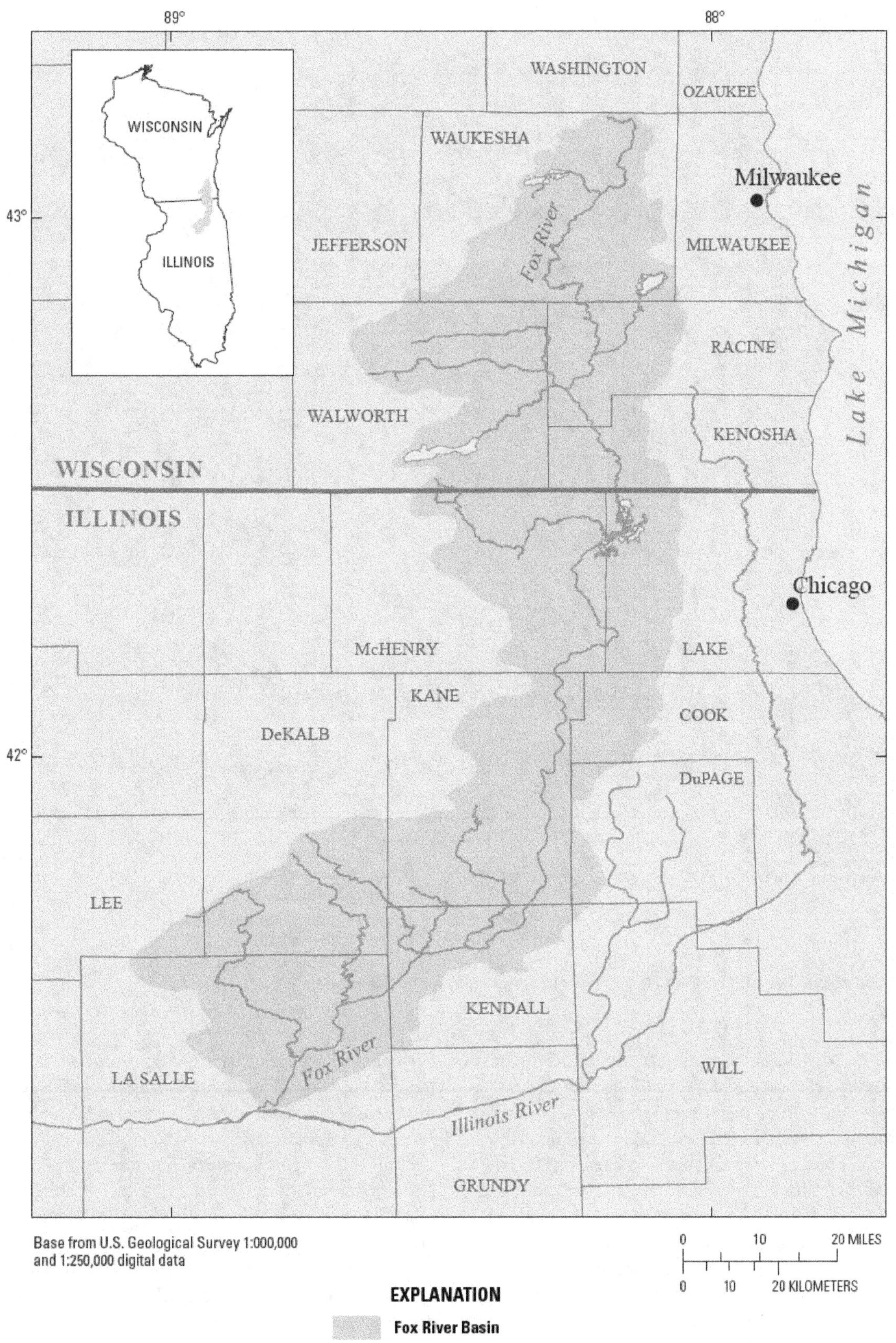

Figure 1*A*. Study area—regional setting.

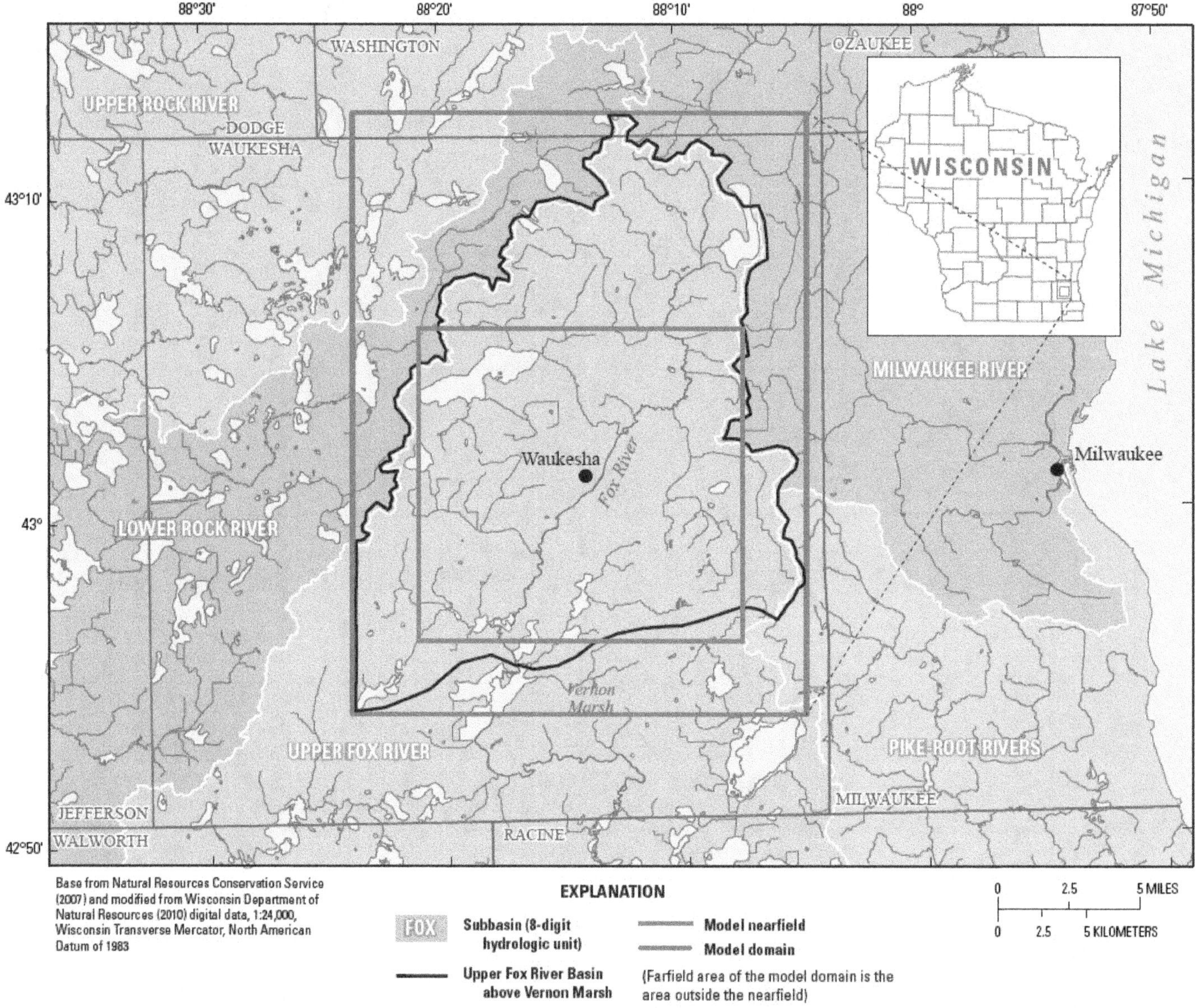

Figure 1*B*. Study area—Upper Fox River Basin, model domain, and model nearfield.

The climatic characteristics of the Upper Fox River Basin can be estimated by reference to data for Waukesha County as a whole. The county receives on average 33.08 inches of precipitation annually, with the highest and lowest monthly rates in June and February, respectively (Southeastern Wisconsin Regional Planning Commission, 2010). Surface-water flow records are reported to indicate that about 7.5 inches of precipitation drain through the surface-water network. The balance infiltrates the soil or is lost by evapotranspiration (Poff and Threinen, 1963). Mean daily temperatures range from a low of 19.1 degrees Fahrenheit in January to a high of 71.9 degrees Fahrenheit in July, averaging 46.9 degrees annually (Southeastern Wisconsin Regional Planning Commission, 2010). Average snowfall is 40.4 inches and soils are commonly frozen during winter months. The 2000 United States census identified 360,767 residents in the county (Southeastern Wisconsin Regional Planning Commission, 2006), which is an increase of 56,052 residents during the 1990s, exhibiting one of the fastest rates of county growth in Wisconsin. Land-use inventories for 2000 showed the land cover to be 36.8 percent urban, 4.5 percent surface water, 14.2 percent wetlands, 7.8 percent woodlands, and 36.7 percent agricultural or open land (Southeastern Wisconsin Regional Planning Commission, 2006).

For convenience, the study area was divided into a *nearfield* (fig. 1*B*), where model discretization and properties are relatively precisely defined, and a surrounding *farfield*, where model inputs are relatively coarsely defined (fig. 1*B*). The model nearfield is centered on the city of Waukesha and encompasses much of the Upper Fox River Basin. The model is designed to simulate groundwater/surface-water interactions at a relatively fine scale in the nearfield.

The vertical extent of the model study area corresponds to the shallow aquifer system in southeastern Wisconsin (Feinstein and others, 2005a). This system is composed of unconsolidated material (glacial and alluvial deposits) overlying Silurian dolomite bedrock (fig. 1*C*). The bottom of the shallow aquifer system corresponds to the contact between the Silurian dolomite and the underlying Maquoketa shale confining unit, or, if the shale is absent, the uppermost Cambrian-Ordovician unit that is present.

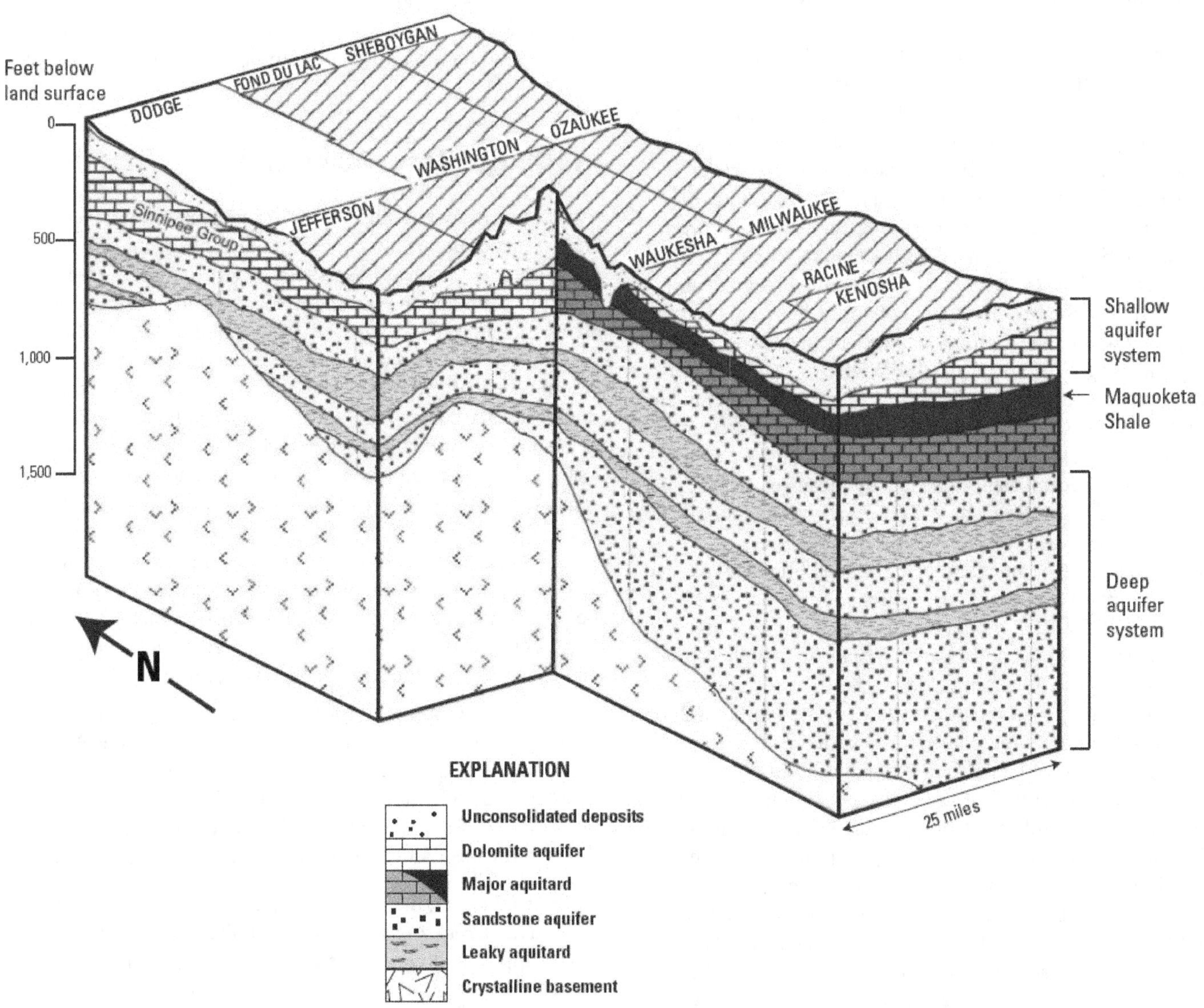

Modified figure from Southeastern Wisconsin Regional Planning Commission, 2002

Figure 1*C*. Schematic diagram showing study area—shallow and deep aquifer systems in southeastern Wisconsin.

1.3 Previous Hydrogeologic Investigations and Modeling Studies

Southeastern Wisconsin (Kenosha, Milwaukee, Ozaukee, Racine, Walworth, Washington and Waukesha Counties) has been the object of a number of studies that focus on groundwater flow. Regional investigations, which incorporated southeastern Wisconsin in a larger multistate framework, include works of the Illinois State Water Survey (Burch, 1991; Meyer and others 2009) and the U.S. Geological Survey (Young and others, 1989; Mandle and Kontis, 1992; Young, 1992; Feinstein and others, 2010). Studies limited to southeastern Wisconsin include efforts restricted to simulating flow in the Cambrian-Ordovician units constituting the region's deep sandstone aquifer (Young, 1976; Jansen and Rao, 1998).

The shallow and deep groundwater resources and geology of the region were investigated in a cooperative effort between the Southeastern Wisconsin Regional Planning Commission (SEWRPC) and the Wisconsin Geological and Natural History Survey (Eaton and others, 1999). This work contributed to the development of a groundwater-flow model for southeastern Wisconsin (the "SEWRPC model"), which integrated the unconsolidated and bedrock aquifers (Feinstein and others, 2005a, 2005b). The SEWRPC model represented all rock units, including confining beds, from land surface to the top of the Precambrian sequence, by means of 18 layers. Minimum grid resolution was 2,500 feet (ft) in the model nearfield in southeastern Wisconsin. The model was calibrated for predevelopment and pumping conditions by using heads and stream base-flow observations for the period 1864 to 2000. The SEWRPC model was used as a tool to evaluate future water supply in southeastern Wisconsin (Southeastern Wisconsin Regional Planning Commission, 2010). In that effort, pumping rates in the model were updated through 2005.

Several hydrogeologic investigations have focused on areas that overlap the Upper Fox River Basin. Stratigraphic and hydraulic data were collected over a 40-mi^2 area along the Fox River in Waukesha County as an aid in evaluating the potential for expanding water supply (Batten and Conlon, 1993). Two recent reports describe groundwater-flow models that focus on the shallow system in parts of Waukesha County (Dunkle, 2008; Jansen and Loughry, 2009). The Dunkle model adopts a geostatistical approach to characterize the unconsolidated sediments based on records from water-well driller logs, whereas, the Jansen and Loughry model interprets the system in terms of stratified units consisting of alternating outwash/ice contact deposits and till/lacustrine deposits. Also available are studies initiated by Waukesha Water Utility along the Upper Fox River that contain withdrawal information, test boring and geophysical data, as well as records of aquifer tests involving pumping of test borings (Aquifer Science and Technology, 2004, 2008, 2010).

2. Conceptual Model

The quantitative analysis of the groundwater-flow system in the Upper Fox River Basin depends largely on the treatment of three elements: the heterogeneity of the subsurface, the exchange between groundwater and surface water, and the stresses on the groundwater system expressed as sources and sinks of water. The objectives particular to this study also require that the model code and model resolution allow precise simulation of the interaction between groundwater and surface water, including the effect of pumping near streams (riparian wells).

2.1 Hydrogeologic Framework

The shallow aquifer system in southeastern Wisconsin consists of unconsolidated sediments overlying Silurian dolomite. The unconsolidated sediments are mostly glacial in origin, although alluvial sediments also are present along stream channels. A key control on local patterns of shallow groundwater flow is the heterogeneity of the glacial sediments. In some parts of the model study area, the sediment texture is predominantly coarse grained or fine grained over the entire unconsolidated thickness, whereas in other areas, repeated episodes of glacial advance and retreat through the Pleistocene Epoch have produced deposits that vary in texture and, therefore, in permeability over relatively short distances.

It is convenient to roughly correlate the degree of heterogeneity in the unconsolidated deposits with the glacial units present in the model study area. The most recent mapping of Pleistocene deposits in Wisconsin (Syverson and others, 2011) defines three glacial units over the area of interest (fig. 2):

- To the west, the *Horicon member* of the Green Bay Lobe was deposited during the last part of the Wisconsin Glaciation. It includes till, associated sand and gravel, and other stratified deposits. Sand content generally is between 60 to 80 percent in this unit, which indicates that the deposits are typically coarse grained.

- To the east, the dominant and most recent unit is the *Oak Creek Formation* of the Lake Michigan Lobe, approximately equivalent in age to the Horicon member (15,000 to 17,000 years old). It consists of fine-grained till with some lacustrine clay, silt, and sand and some glaciofluvial sand and gravel. Generally, the till texture is silty clay or silty clay loam, composed of 80 to 90 percent silt. This unit is largely fine grained.

Modified from Syverson and others, 2011

Figure 2. Distribution of surficial Pleistocene lithostratigrahic units in Wisconsin.

- In the middle of the study area and extending over most of the Upper Fox River Basin, the unconsolidated deposits correspond to the *New Berlin Member* of the Lake Michigan Lobe, equivalent in age to the Oak Creek Formation. The New Berlin Member is reported to consist of two principal facies, a sand and gravel unit typically overlying a till unit (Syverson and others, 2011). However, comparison of available water-well driller logs and geologic logs suggests that unlike the area to the west where deposits are predominantly coarse grained and the area to the east where deposits are predominantly fine grained, the area associated with the New Berlin Member shows little continuity between sand/gravel and silt/clay deposits. A typical stratigraphic section through the unconsolidated material based on the evidence of subsurface logs shows a high degree of heterogeneity. Figure 3 contains a schematic section, which represents the short distances over which the textures of deposits vary as well as the overall lack of continuous layering. This scale of variation can be attributed to a complicated history of glacial advances, which deposited mostly clayey till, interrupted by melting, which produced meltwater channels that initially eroded the fine-grained material and later deposited sandier material. Stacking of erosional and depositional phases would tend to interrupt the continuity of clayey till sheets and sandy channel deposits (Professor Douglas Cherkauer, University of Wisconsin-Milwaukee, oral commun., April 2, 2011).

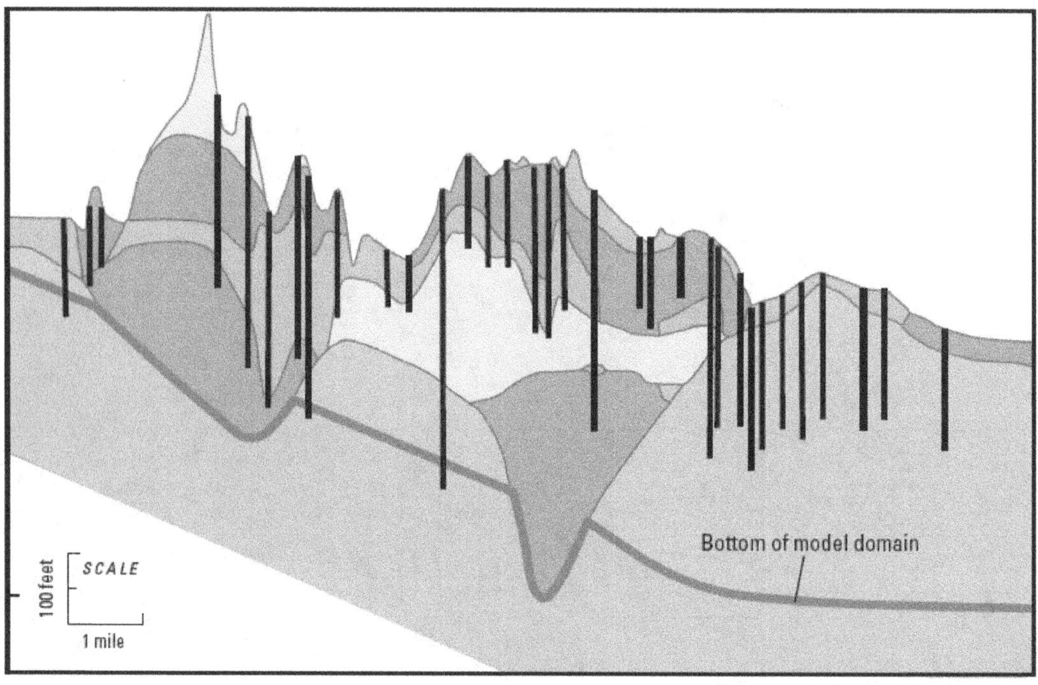

Schematic modified from D.S. Cherkauer, Geosciences Department, University of Wisconsin–Milwaukee

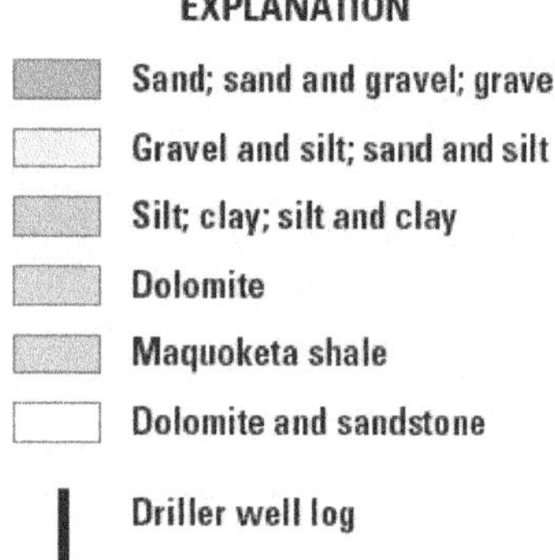

Figure 3. Schematic hydrogeology showing glacial heterogeneity.

The "Model Construction" section of this report presents a stratigraphic database for the model domain based on information from available logs. The irreducible uncertainty in mapping the texture of the unconsolidated deposits because of the spacing and vertical penetration of logs, as well as the difference in quality and reliability among logs, makes it difficult to definitively assess the degree of continuity of coarse-grained and fine-grained deposits, especially in the parts of the Upper Fox River Basin dominated by the New Berlin Member. For this reason, the database is used to generate two interpretations of the subsurface—one favors the connection among fine-grained deposits and one favors the connection among coarse-grained deposits. Of course, neither interpretation is a "true" representation of the unconsolidated heterogeneity. The intent is to construct and calibrate two Upper Fox River Basin groundwater-flow models from the subsurface data in an effort to at least partly encompass the range of uncertainty in the characterization of the glacial sediments. In this way, different subsurface conditions, such as preferential flow paths associated with relatively continuous channel deposits and confined conditions associated with layering of till sheets, can be represented and their effect on groundwater-flow patterns (for example, on the sources of water to wells) can be compared. The model that preferentially connects fine-favored deposits is called the *fine-favored model*; the model that preferentially connects coarse-favored deposits is called the *coarse-favored model*.

In addition to marked heterogeneity, the unconsolidated deposits display variable thickness across the study area (fig. 4*A*). The deposits are present over more than 99 percent of the model domain (in a few places the underlying dolomite is at the surface) and is at least 100 ft thick in 50 percent of the model domain but only in 34 percent of the model nearfield. The thickness is an important control on the transmissivity of the glacial deposits and their ability to support pumping. However, some zones of greater than average thickness (for example, the bedrock valley in the southern part of the study area, located in the farfield of the model domain [fig. 4*A*]) do not necessarily correlate with the zones of greatest well yields, which also depend on the texture of the deposit and the potential to induce water from streams (Batten and Conlon, 1993).

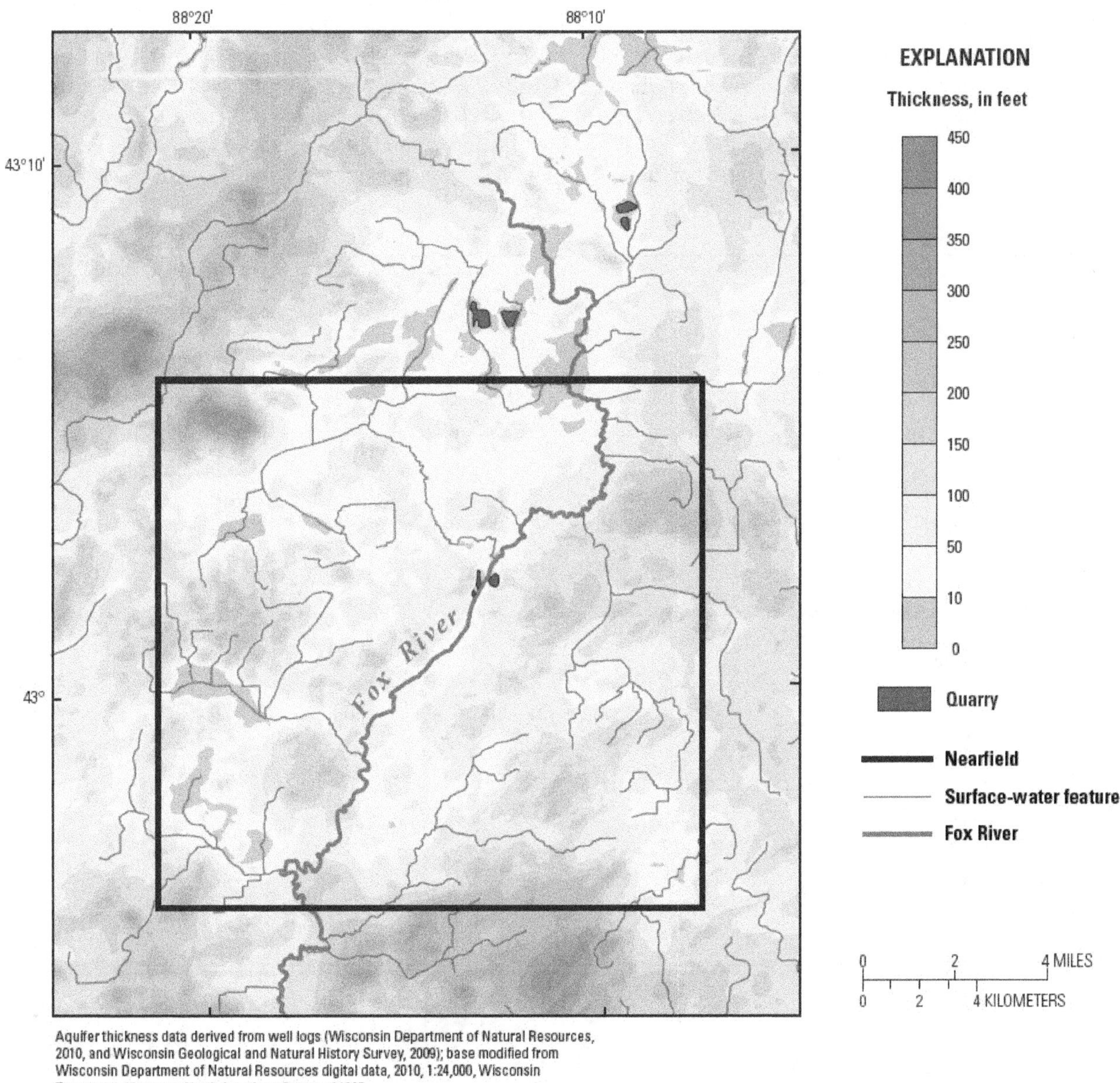

Figure 4*A*. Thickness (feet) of shallow aquifer system in model domain—unconsolidated deposits.

The Silurian dolomite is the top of bedrock over most of the study area, thickening to the east (fig 4*B*). It is an important aquifer in southeastern Wisconsin, supplying many domestic and some public and industrial wells (Southeastern Wisconsin Regional Planning Commission, 2010). As is typical for carbonate rock in the Upper Midwest, the ability of the Silurian dolomite to transmit water is related to the properties of the matrix and to the weathering and development of fractures. For instance, numerous investigators have reported higher than background matrix hydraulic conductivity values in a weathered zone near the carbonate bedrock surface (Carlson, 2001; Stocks, 1998) or in zones of bedding-plane (Eaton, 2002; Muldoon and others, 2001) and vertical fractures (Jansen, 1995). Following the approach adopted in the SEWRPC regional model (Feinstein and others, 2005a) based on field studies in southeastern Wisconsin (Rovey, 1990), this study assumes that fractures and dissolution because of weathering enhance the hydraulic conductivity in the upper 20 ft of the Silurian dolomite thickness.

The bottom boundary of the shallow aquifer system is the top of the Maquoketa shale (fig. 3), a confining unit that inhibits vertical exchange with the deep aquifer system consisting of Cambrian-Ordovician rocks (Eaton, 2002). However, the large rate of withdrawals from the deep aquifer system (Feinstein and others, 2005a) have created a regional cone of depression at depth that induces some water to leak from the shallow to the deep parts of the flow system (Feinstein and others, 2005b). Account must be taken of this vertical flux in evaluating groundwater conditions for the Upper Fox River Basin.

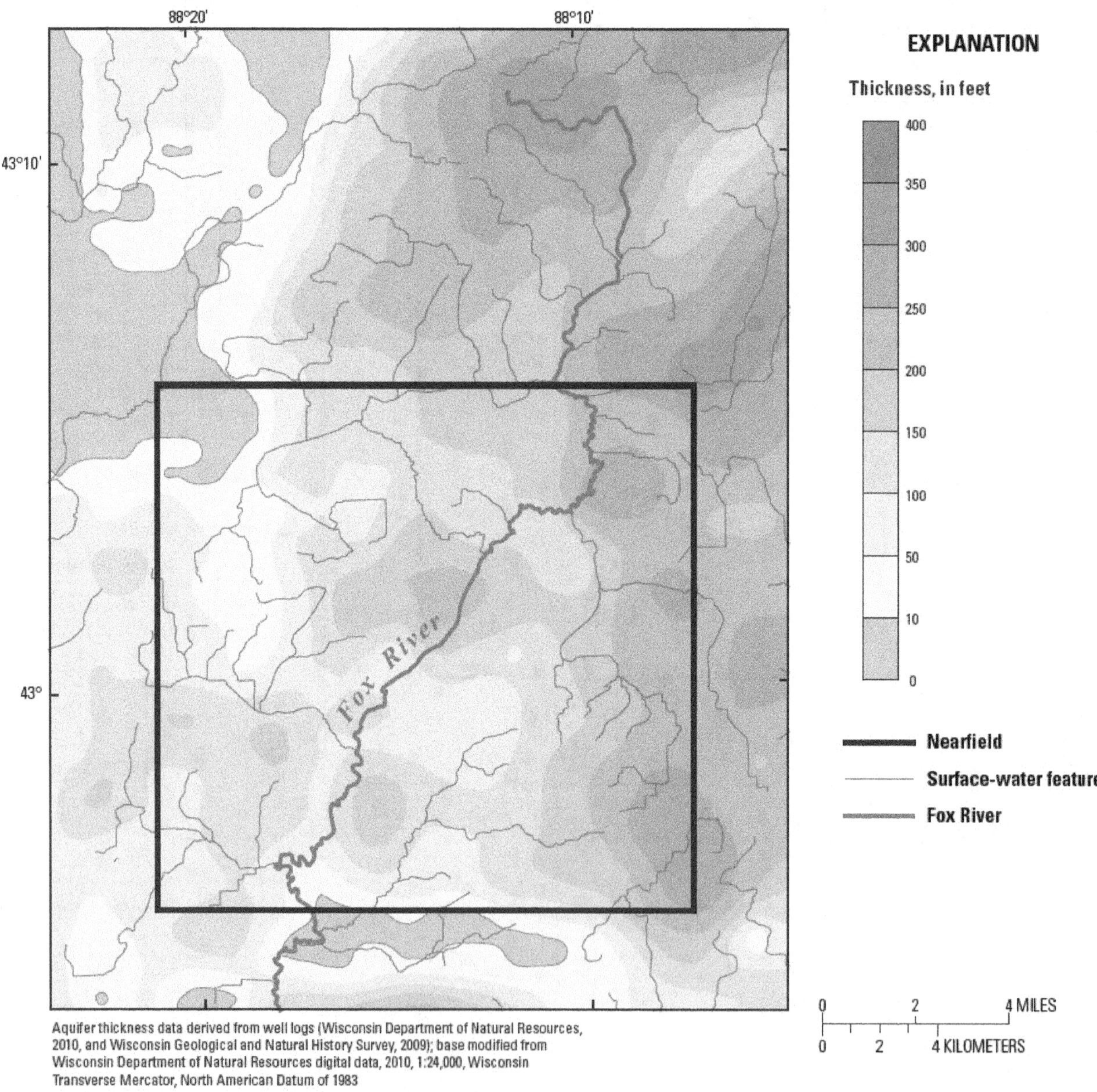

Figure 4*B*. Thickness (feet) of shallow aquifer system in model domain—Silurian dolomite.

2.2 Hydrologic Framework

The water-table elevation and water levels at depth in the shallow aquifer system in southeastern Wisconsin are strongly influenced by the undulations of the land-surface topography. The most direct control on water levels is the surface-water network, which tends to fix the water-table elevation along valleys and lowlands (figs. 5*A* and 5*B*). The surface-water network in the model domain consists of the Fox River, tributary streams, and 22 water bodies (lakes, ponds and wetlands) typically connected to streams (fig. 6). The area also contains riparian wetlands which drain to streams and lakes as well as upland wetlands which are less likely to be integrated in the surface-water network. The Vernon Marsh at the south end of the Upper Fox River Basin is a complex of wetlands whose water level is partly controlled by a system of weirs (operated by the Wisconsin Department of Natural Resources) on tributaries to the Marsh such as Pebble Creek and Mill Brook (Professor Douglas Cherkauer, University of Wisconsin-Milwaukee, oral commun., September 2009).

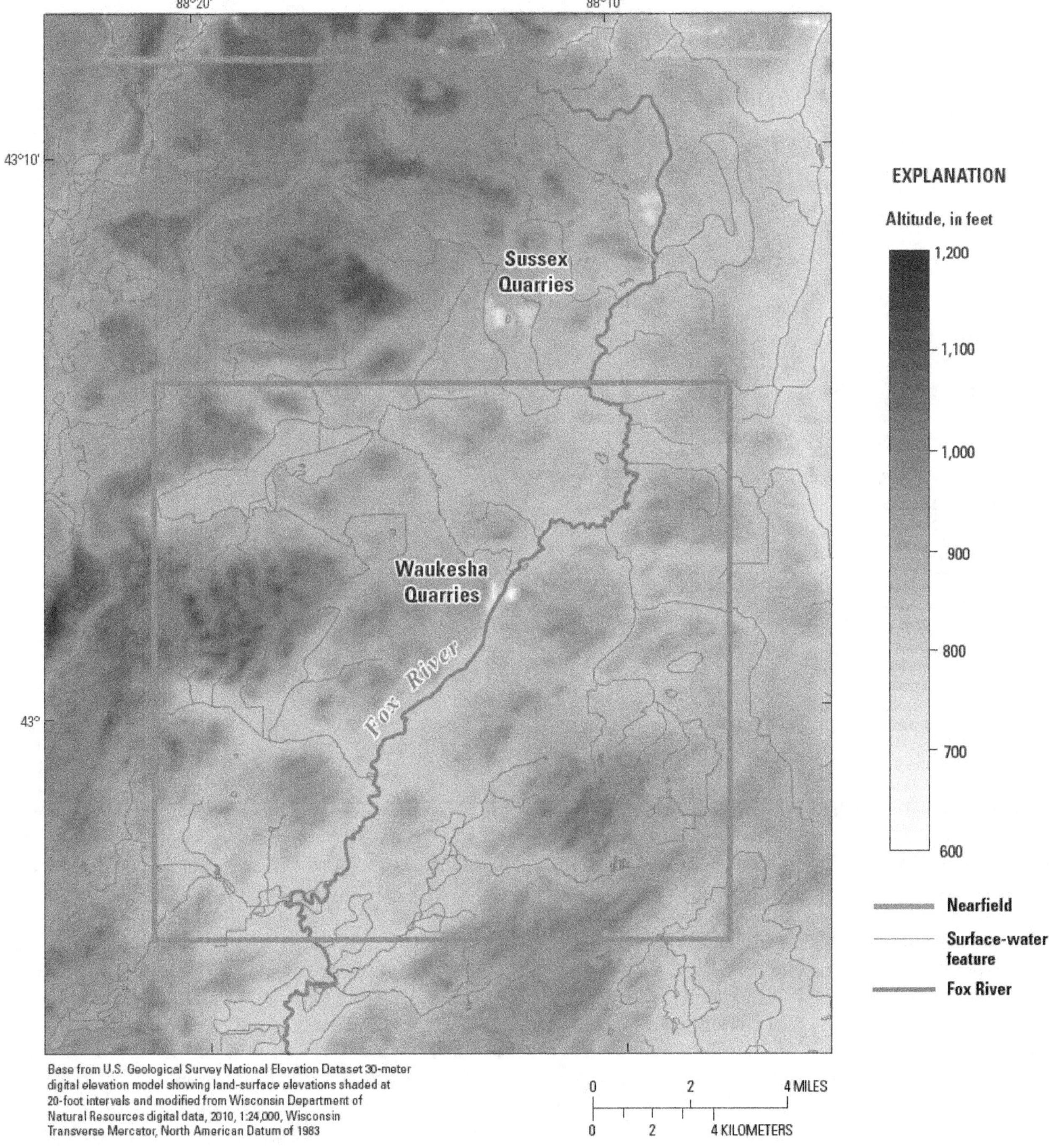

Figure 5*A*. Land-surface elevation—topography in model domain.

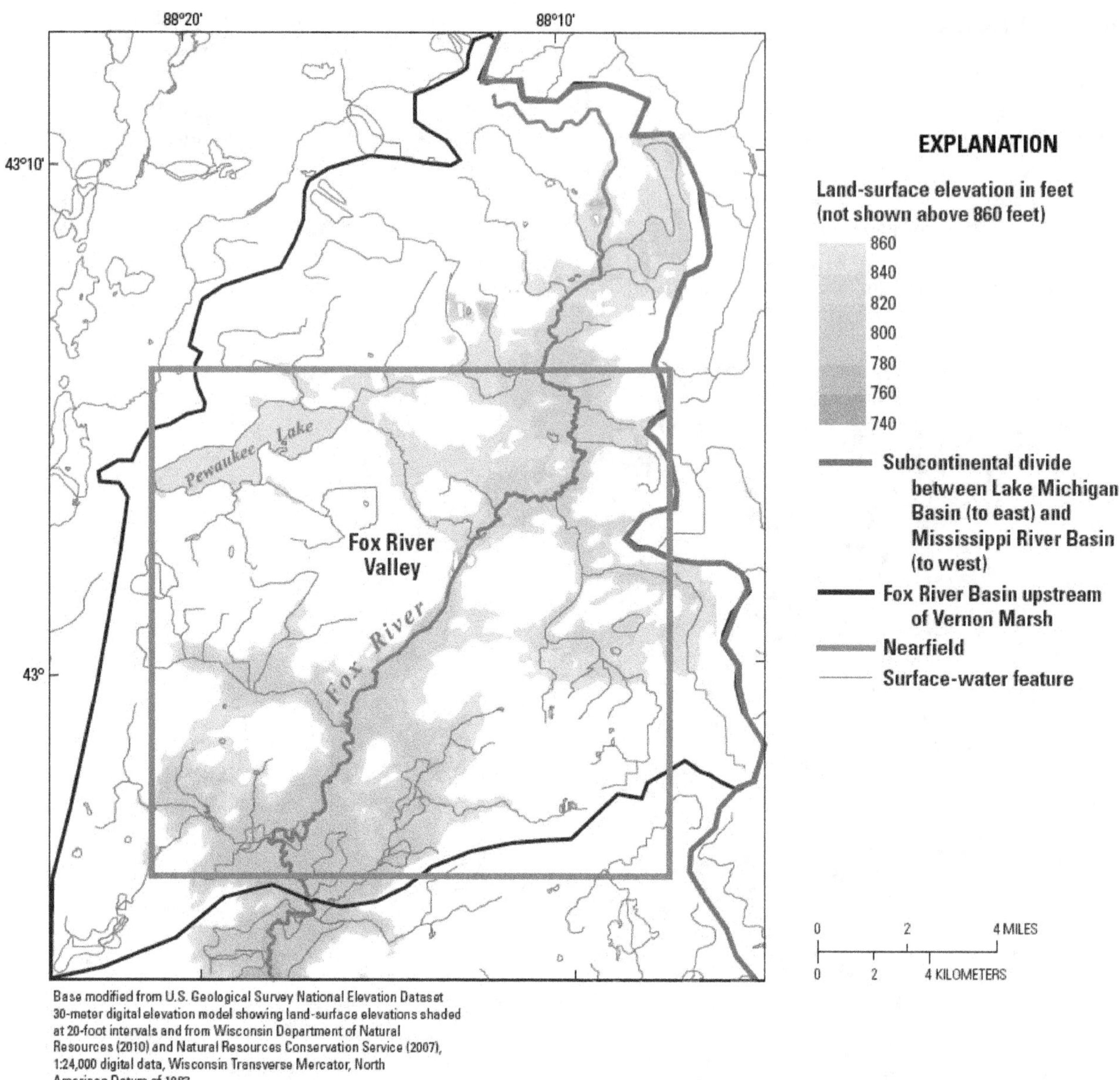

Figure 5*B*. Land-surface elevation—Fox River valley and tributary lowlands.

The main tributaries to the Upper Fox River are Lannon Creek and Sussex Creek in northern Waukesha County; Pewaukee River and Poplar Creek immediately north of the city of Waukesha; and Pebble Creek, Genesee Creek and Pebble Brook between the city of Waukesha and the Vernon Marsh (fig. 6*A*). Pewaukee Lake is a large feature that drains to the Pewaukee River and then on to the Fox River. A chain of lakes is present in the model farfield in the Bark River and Oconomowoc River Basins located northwest of the Fox River Basin. Other farfield water bodies include Tamarack Swamp near the headwater of the Fox River, and Spring Lake and Little Muskego Lake south of the Upper Fox River Basin (fig. 6*A*).

A primary objective of the Upper Fox River Basin models is to simulate base flow to the surface-water network. For this study, base flow is defined not only to include the groundwater contributions to streams and lakes, but also the effluent discharge from wastewater-treatment plants (WWTPs). The plants in the model domain recirculate water withdrawn from shallow and deep wells (chiefly public supply wells for the main population centers) to the Fox River. There are three WWTPs that discharge to the Fox River, which serve the Cities of Sussex, Brookfield, and Waukesha (fig. 6*B*). Records indicate that the effluent flow from these plants has increased over the last 50 years but that in any given year the effluent flow is relatively steady. Records also indicate that the combined discharge is not an inconsequential fraction of the Fox River streamflow during low-flow periods. The WWTP flux is discussed further in the "Model Construction" section.

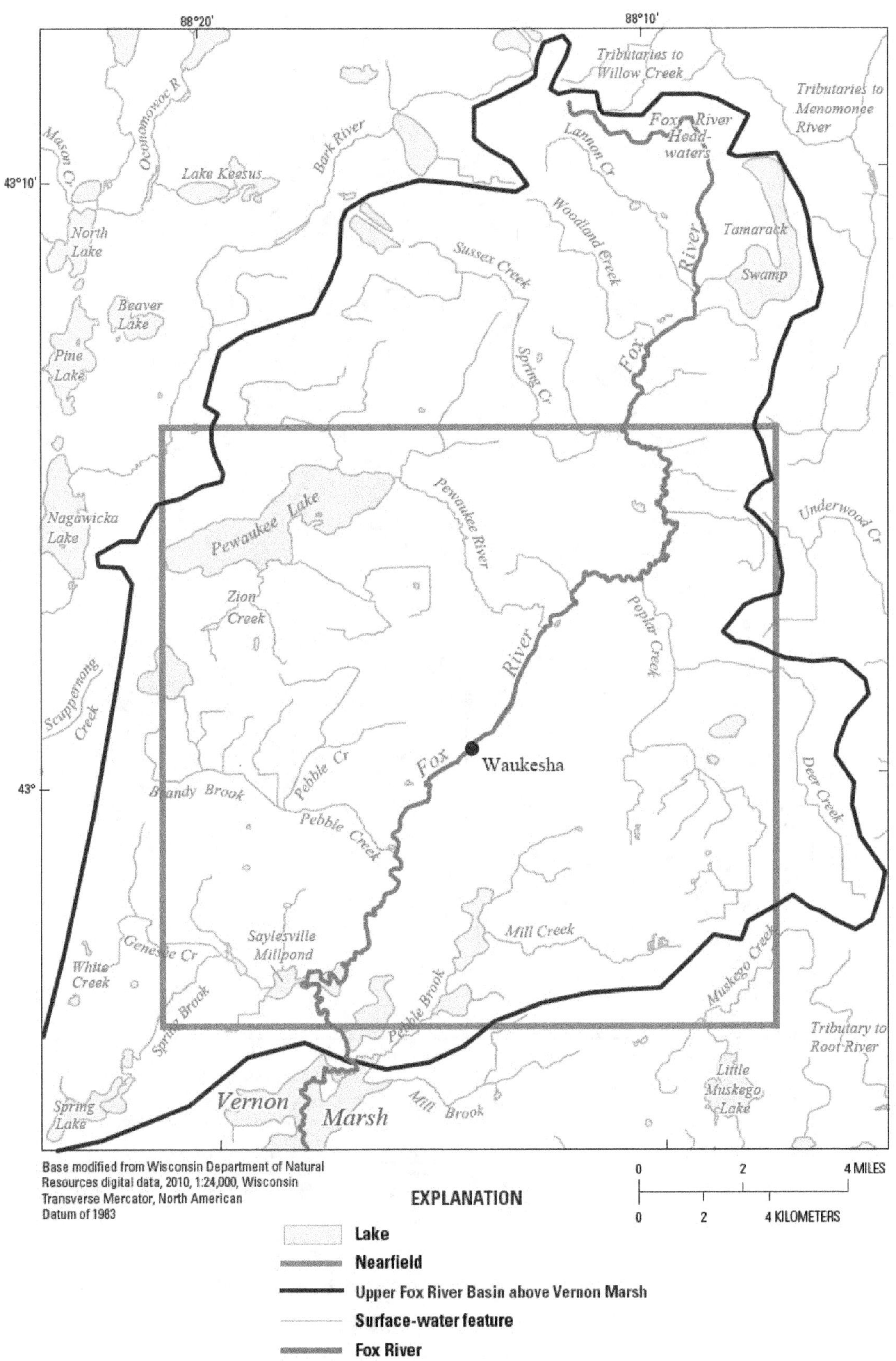

Figure 6A. Surface-water network in model domain—streams, lakes, and wetlands.

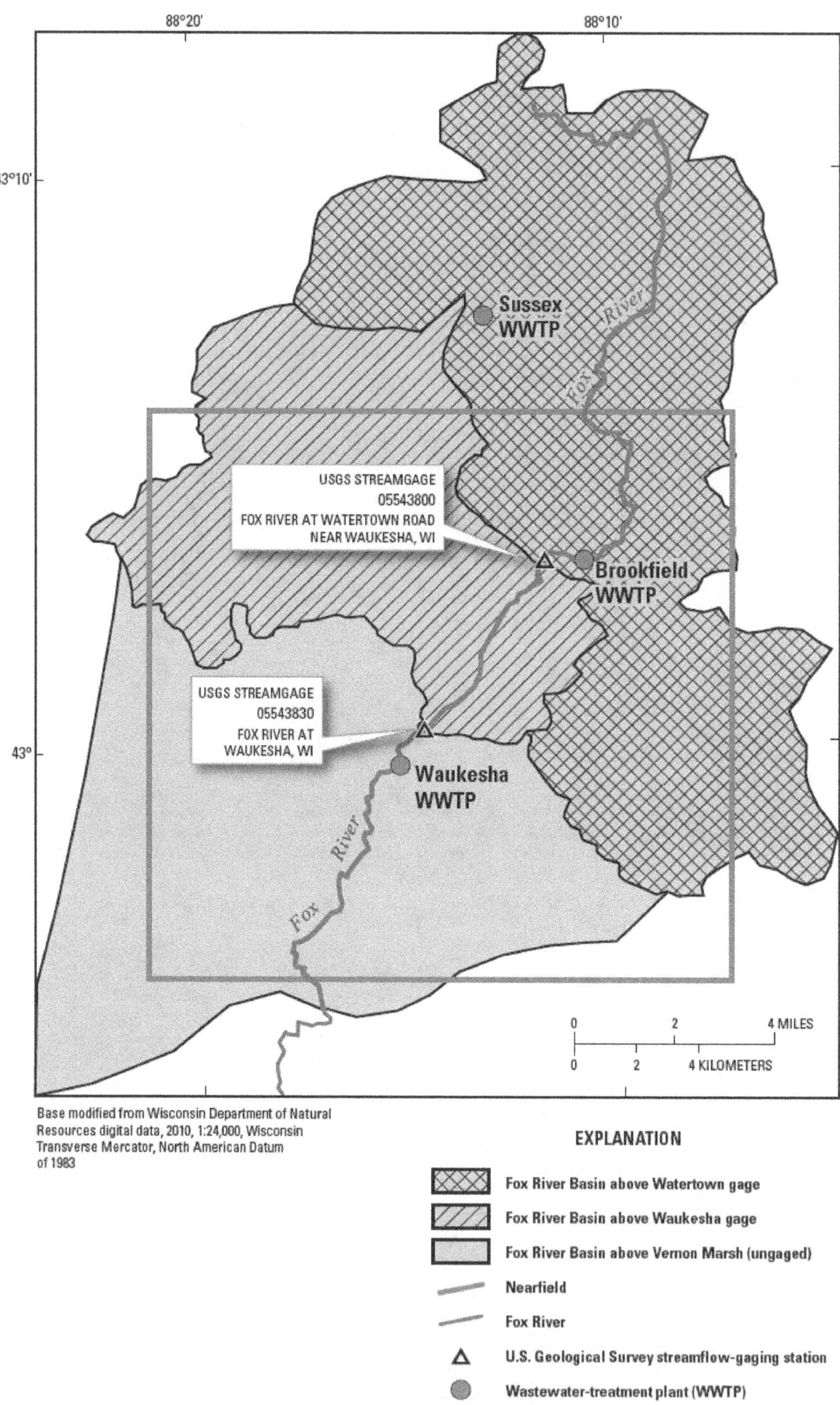

Figure 6*B*. Surface-water network in model domain—streamgages, wastewater-treatment plants, and nested drainage basins.

There are two U.S. Geological Survey (USGS) gages on the Upper Fox River (fig. *6B*), one located at Watertown (gage 05543800, active from 1992 to 2000) and one located at Waukesha (gage 05543830, active from 1963 to present). Streamflow records from these gages have been used to estimate long-term average rates of base flow (groundwater contribution plus effluent discharge) for their respective subbasins within the Upper Fox River Basin. The estimation and use of the base-flow values are described in the "Model Calibration" section.

2.3 Sources and Sinks of Water

The stresses on the Upper Fox River Basin groundwater-flow system consist of fluxes into the model domain (sources) and out of the model domain (sinks). The sources of water are:

- the fraction of precipitation that percolates as recharge to the water table,
- losses from surface-water bodies to the subsurface, and
- inflows across the lateral and vertical domain boundaries.

The sinks for water are:

- discharge to surface-water bodies,
- outflows across the lateral and vertical domain boundaries,
- withdrawal from pumping wells, and
- discharge to quarries excavated at the top of the Silurian dolomite.

The evaluation of these flux terms is based on several sources: recorded data, estimation techniques, previous modeling at the regional scale, and outputs of the model itself. The evaluation, location, and zonation of the fluxes input to the model are discussed in the "Model Construction" section.

An additional potential source/sink for the groundwater-flow system is the release of water from or addition of water to aquifer storage under transient (time-varying) conditions. In this application, the storage flux is only considered as part of the calibration process (see "Model Calibration" section).

2.4 Code Selection and Model Resolution

The Upper Fox River Basin models are designed to simulate interactions between the shallow groundwater-flow system and the surface-water system at a scale sufficiently refined to (1) include and integrate virtually all individual elements of the surface-water network and (2) simulate local exchanges between the subsurface and surface systems, including flow induced from streams to riparian wells. These objectives require the use of advanced modeling packages and special attention to the discretization of the model grid.

In this study, the groundwater-flow system is simulated using packages that are part of the USGS MODFLOW-2005 numerical code (Harbaugh, 2005), a finite-difference program supporting three-dimensional and transient solutions. The MODFLOW-2005 code features several advanced packages applied in this study for linking groundwater to surface water and for routing water through the surface-water system. The streams, lakes, and other water bodies in the Upper Fox River surface-water network are simulated by means of the Streamflow-Routing (SFR) and Lake (LAK) packages as a single integrated system, which routes water from headwater features through higher-order features and to the outlet of the Upper Fox River Basin at Vernon Marsh. Surface-water features outside the Upper Fox River Basin (all located in the model farfield) are represented using simpler MODFLOW algorithms—the River (RIV) and Drain (DRN) packages—which simulate groundwater/surface-water interactions but do not route water through streams and water bodies or calculate surface-water stages. Other MODFLOW-2005 packages employed to represent fluxes include RCH for recharge and WEL for pumping wells and boundary inflows and outflows.

The degree of accuracy attained in numerical models that simulate groundwater/surface-water interactions is highly dependent on the horizontal and vertical grid resolution (Haitjema and others, 2001; Hunt and others, 2003; Haitjema and others, 2010). A discretization that is too coarse tends to group multiple surface-water features in a single model cell, which obscures the role of individual streams and lakes. Moreover, the numerical code assigns the properties of the surface-water features (the stage and bed conductance) to the spatial center of the model cell, thereby possibly mislocating the origin of the stress. When wells are located near surface-water bodies, the spatial relation between the features can be distorted unless the grid is fine enough both laterally and vertically so that the cell centers (where the well withdrawal is simulated) are at approximately the correct spatial distances.

A related set of problems involves errors in modeled groundwater/surface-water exchanges that may result from an inaccurate representation of the *leakage between model layers* in the presence of surface-water features inserted in coarse grid cells. One effect of a relatively coarse grid is to smear the pattern of upward groundwater discharge to streams over too large an area; another is to miscalculate the magnitude of the discharge (Haitjema and others, 2010; Feinstein and others, 2010, appendix 2). To understand these issues it is helpful to consider the "leakage factor" λ (feet) (Hantush and Jacob, 1954; Verruijt, 1970), also referred to as the "characteristic leakage length" (Haitjema and others, 2001; Bakker and Strack, 2003). For a two-layer aquifer system, the characteristic leakage length is defined as:

$$\lambda \quad \sqrt{\frac{T_u T_l c}{T_u + T_l}} \tag{1}$$

where

T_u and T_l are the transmissivities of the upper and lower layer, respectively, in feet squared per day, and

c is the vertical resistance (thickness divided by vertical hydraulic conductivity) of a separating confining unit or the aquifer material, in days.

The characteristic leakage length determines whether the vertical leakage to or from streams or wells is concentrated locally or is distributed ("smeared out") over a larger area. For example, it can be shown that the leakage induced by a stream nearly vanishes beyond a distance of 3λ from the stream boundary (Hunt and others, 2003). Thus, for small λ-values, the leakage is concentrated near the stream, whereas for large λ-values, it occurs over a much larger area. Haitjema and others (2001) found that in order to obtain an acceptable representation of the leakage distribution in finite-difference models, such as MODFLOW-2005, the cell size should be less than λ, preferably as low as 0.1λ. When the model cell size is larger than this threshold value, inaccurate leakage distributions may result in inaccurate simulation of heads and streamflows.

Application of the leakage factor to transmissivity and vertical resistance values typical for the shallow system in southeastern Wisconsin serves as a guideline for the proper grid resolution. Assuming that horizontal hydraulic conductivity is 5 feet per day (ft/d) and vertical hydraulic conductivity is 0.01 ft/d for unconsolidated and bedrock units (reasonable values for sandy till and dolomite), and the average thicknesses are 115 ft and 50 ft for the unconsolidated and bedrock units (typical values in Waukesha County) transmissivity values are 575 and 250 ft^2/d, and the resistance value for the vertical thickness between cell centers is 8,250 days. The resulting leakage length, l, is about 1,200 ft (Feinstein and others 2010, appendix 2). This result implies that a lateral grid spacing on the order of 0.1l or about 120 ft will ensure that the grid resolution does not introduce inaccuracies into the solution. The nearfield grid resolution for the Upper Fox River Basin models is 125 ft per side, which should be small enough to virtually exclude numerical distortions. The fine grid spacing has the added advantage of allowing relatively precise location of features such as streams and wells, including riparian wells.

The vertical layer spacing is also made relatively fine in the Upper Fox River Basin models in order to allow accurate input of partially penetrating wells and also to allow accurate simulation of vertical gradients especially in the vicinity of surface water. The top layer of the model—designed to host the boundary conditions corresponding to surface-water features—extends from the land surface to a depth of only 20 ft. (See "Model Construction" section for full explanation of layering scheme.) Static water levels from the database of water-well driller logs for shallow wells in the study area suggest that the median water-table elevation is almost 40 ft below the land surface, with depths approaching zero in the valleys and depths typically greater than 40 ft under the uplands. (See "Model Calibration" section for an account of available water-level information.) Accordingly, the models are expected to simulate dewatered conditions for the top layer over a large fraction of the model domain area. Model layers representing deeper parts of the groundwater system are expected to be dewatered under upland areas. MODFLOW-2005 allows dewatering of model cells, but the simulation of water-table conditions below the top model layer can introduce instabilities to the model solution as well as unwanted sensitivity to initial conditions. Use of a recently developed version of MODFLOW-2005, called MODFLOW-NWT (Niswonger and others, 2011), offers a way to overcome the difficulties associated with thin model layers under unconfined conditions. The "Model Solver" section of this report describes in more detail the motivation for and the implementation of the NWT formulation to this project.

The final element of the conceptual model involves the temporal resolution of the model solutions. The base solutions are steady-state simulations and reflect groundwater conditions in equilibrium with the source inflows and sink outflows described above. Consideration of transient effects in the development of the fine-favored and coarse-favored models is limited to the calibration phase (the "Model Calibration" section) involving simulation of an aquifer test.

As discussed in the "Model Construction" section, the well withdrawals correspond to 2005 conditions. However, other fluxes associated with the model, such as edge boundary fluxes, recharge to the basin, Pewaukee Lake outflow, and wastewater-treatment effluent to the Fox River, are referenced to separate time frames—that is, long-term average conditions in the case of recharge or post 2005 conditions in the case of lake outflow and effluent. In this sense, the output of the models is best characterized as reflecting conditions in equilibrium with ***recent*** flow rates from sources and sinks. Possible errors deriving from the approximate nature of the flux estimates and the neglect of transient effects corresponding to adjustment of the flow system to these time-dependent boundary conditions are discussed in the "Model Limitations" section of the report.

3. Model Construction

The model construction involves the horizontal and vertical discretization of the model domain into finite-difference cells along with the definition of boundary conditions at the edge of the grid, the translation of the unconsolidated and bedrock hydrogeology into hydraulic conductivity zones, the distribution of water-table recharge acting as the principal source of water to the system, the representation of the principal sinks for groundwater corresponding to various surface-water features, and the distribution of local groundwater sinks associated with pumping wells and with flow to dolomite quarries. Most elements of the model construction are common to the fine-favored and coarse-favored model—the only element that differs between the two is the representation of the hydraulic conductivity zones.

The platform used to help construct the flow models (as well as to visualize input and output) was Groundwater Vistas, Version 5 (Rumbaugh and Rumbaugh, 2007).

3.1 Model Grid

The model grid consists of the lateral discretization of the model domain into rows and columns and the vertical discretization of the unconsolidated deposits and the Silurian dolomite into layers. At the lateral and bottom edges of the grid, groundwater enters and exits the model domain as a function of prescribed fluxes.

3.1.1 Lateral Discretization

The MODFLOW finite-difference grid consists of 508 rows from north to south extending 21.3 mi and 508 columns from west to east extending 16.1 mi (fig. 7), yielding a total area of 343 mi^2. The spacing of the rows and columns is nonuniform. The rows and columns at the edge of the domain are 2,500 ft in width. The grid spacing is reduced away from the edges of the model domain at a ratio of approximately 1.4 until the nearfield is reached, which roughly coincides with the north/south extent between the confluences of Sussex Creek and Pebble Brook with the Fox River (about 11.3 mi) and the west/east extent between the west side of Pewaukee Lake and the east edge of the Upper Fox River Basin (about 11.6 mi). Inside the nearfield the row and column widths are uniformly set to 125 ft per side. The nearfield grid corresponds to model rows 22 to 498 and to model columns 11 to 498 and covers an area of 131 mi^2.

3.1.2 Vertical Layering

The model is divided into seven layers. Layers 1 through 5 represent unconsolidated material and layers 6 through 7 represent the Silurian dolomite. However, not all unconsolidated layers necessarily participate from a flow standpoint at a given row/column location. The number of participating layers depends on the total unconsolidated thickness. For example, if the unconsolidated material is 10 ft thick at a location, then only layer 1 transmits flow whereas layers 2 through 5 are considered "pinched"; each is assigned a minimal thickness equal to 0.5 ft. In this case, the properties of the pinched layers 2 through 5 correspond to the properties assigned the active cell in layer 1. As a second example, if the unconsolidated material is 30 ft thick, then layers 1 and 2 transmit flow whereas layers 3 through 5 are "pinched", and each are assigned a thickness of 0.5 ft and the properties of the overlying layer 2. All five layers are active only if the unconsolidated material is more than 150 ft thick. The explicit logic for determining the number of active unconsolidated layers at a location is presented in table 1. The logic for assigning layers to the Silurian dolomite is simpler—the upper 20 ft, associated with a weathered interval, correspond to layer 6 and any additional thickness is assigned to layer 7.

An important element of the Upper Fox River Basin model is the use of the land-surface elevation as a datum for the delineation of model layer bottom elevations as a function of specific depths. That is, the bottom of layer 1 corresponds to a depth of 20 ft below the land surface or, if the unconsolidated material is less than 20 ft, to the bottom of the unconsolidated material. Analogously, the bottoms of layers 2, 3, and 4 are associated with depths of 50 ft, 100 ft, and 150 ft, respectively, or to the bottom of the unconsolidated material. The bottom of layer 5 corresponds to the bottom of the unconsolidated material for locations where the unconsolidated material is more than 150 ft thick—otherwise the layer bottom corresponds to the bottom of unconsolidated deposits adjusted slightly to take account of the nominal thickness of pinched cells. This logic implies that the maximum thickness of the layers is 20 ft for layer 1, 30 ft for layer 2, 50 ft for layer 3, and 50 ft for layer 4. The layering, which results from the use of the land surface as the datum, is shown for an example model row in figure 8.

The land-surface elevation for the model (fig. 5*A*) was derived from digital terrain models (DTM) of the area. Elevation data for the part of the model domain inside Waukesha County (95 percent of the grid area and 100 percent of the nearfield) were extracted from Waukesha County 2005 DTM files published by Aero-Metric, Inc., Sheboygan, Wisconsin, on behalf of Waukesha County Department of Parks and Land Use (Waukesha County, 2005). These files are based on a Light Detection and Ranging (LIDAR) coverage of the county at an extremely fine (feet scale) resolution. The DTM files smooth the LIDAR data at a 1-inch-equals-100-ft scale to support land-surface mapping at a 2-ft contour interval. Elevation data for Washington County at the north edge of the grid were based on 30 m (approximately 100 ft) digital elevation model data obtained from the U.S. Geological Survey (2010a).

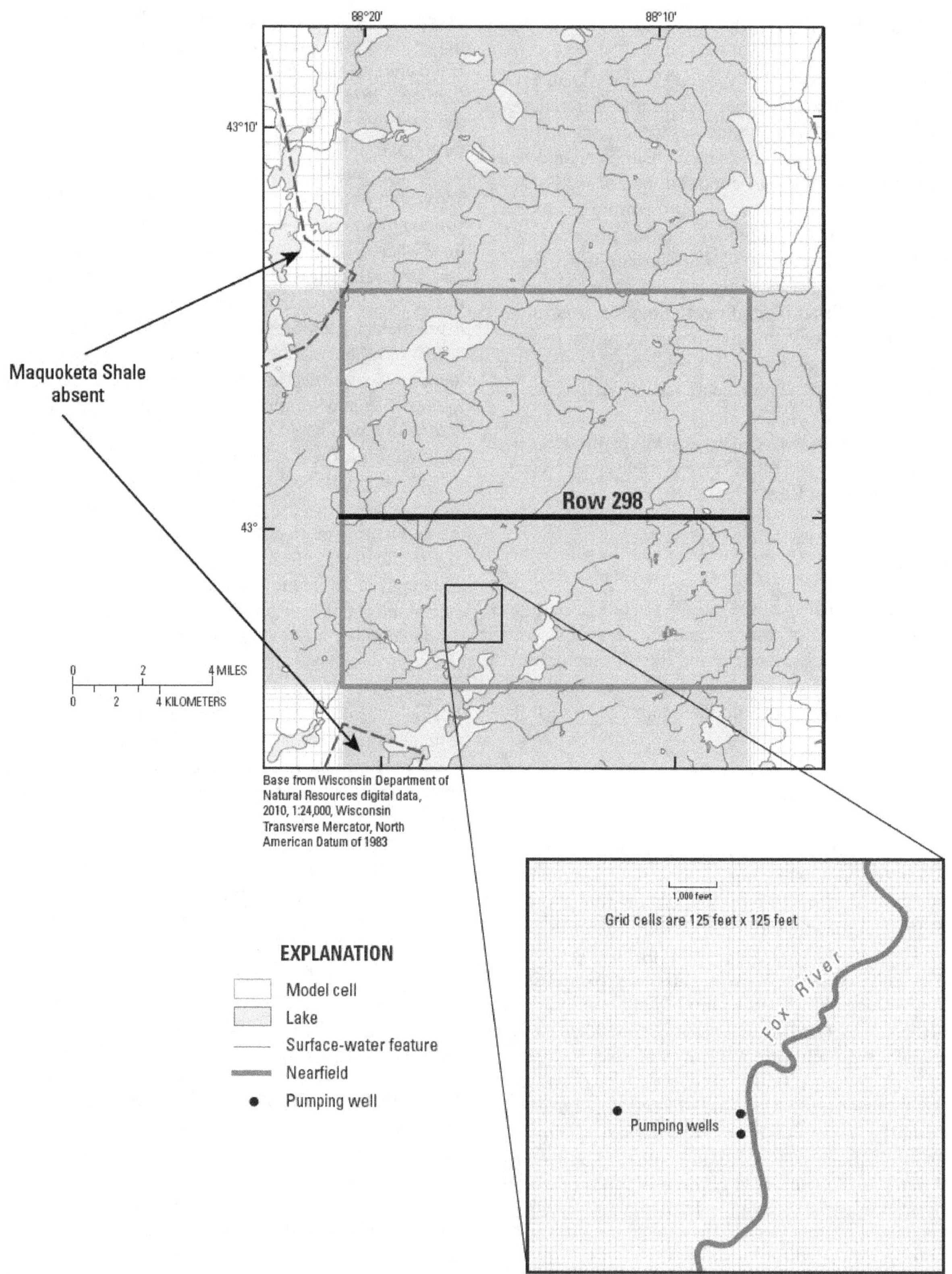

Figure 7. Model grid with nearfield zoom.

Table 1. Layering logic at cell locations for groundwater-flow model of Upper Fox River Basin, southeastern Wisconsin.

Layer 1 (unconsolidated)

- If unconsolidated deposits are greater than 20 feet (ft) thick, bottom elevation of layer 1 is set 20 ft below land-surface elevation;
- If unconsolidated deposits are less than 20 ft thick but greater than 0 ft thick, bottom elevation of layer 1 is set to land surface less unconsolidated deposit thickness;
- If unconsolidated deposits are absent, the layer is "pinched" at this location and the bottom elevation of layer 1 is set to 0.5 ft below land-surface elevation.

Layer 2 (unconsolidated)

- If unconsolidated deposits are greater than 50 ft thick, bottom of layer 2 is set 50 ft below land-surface elevation;
- If unconsolidated deposits are less than 50 ft but greater than 20 ft thick, bottom of layer 2 is set to land-surface elevation less unconsolidated deposit thickness;
- If unconsolidated deposits are less than or equal to 20 ft thick, the layer is "pinched" at this location and the bottom of layer 2 is set to 0.5 ft below bottom elevation of layer 1.

Layer 3 (unconsolidated)

- If unconsolidated deposits are greater than 100 ft thick, bottom elevation of layer 3 is set 100 ft below land-surface elevation;
- If unconsolidated deposits are less than 100 ft but greater than 50 ft thick, bottom elevation of layer 3 is set to land-surface elevation less unconsolidated deposit thickness;
- If unconsolidated deposits are less than or equal to 50 ft thick, the layer is "pinched" at this location and the bottom elevation of layer 3 is set to 0.5 ft below bottom elevation of layer 2.

Layer 4 (unconsolidated)

- If unconsolidated deposits are greater than 150 ft thick, bottom elevation of layer 4 is set 150 ft below land-surface elevation;
- If unconsolidated deposits are less than 150 ft but greater than 100 ft thick, bottom elevation of layer 4 is set to land-surface elevation less unconsolidated deposit thickness;
- If unconsolidated deposits are less than or equal to 100 ft thick, the layer is "pinched" at this locations and the bottom elevation of layer 4 is set to 0.5 ft below bottom elevation of layer 3.

Layer 5 (unconsolidated)

- If unconsolidated deposits are greater than 150 ft thick, bottom elevation of layer 5 is set to land-surface elevation less unconsolidated deposit thickness;
- If unconsolidated deposits are less than 150 ft thick, the layer is "pinched" at this location and the bottom elevation of layer 5 is set to 0.5 ft below bottom elevation of layer 4.

Layer 6 (Silurian dolomite, assumed weathered zone)

- If dolomite thickness is greater than 20 ft, bottom elevation of layer 6 is set to 20 ft below bottom elevation of unconsolidated deposits;
- If dolomite is less than 20 ft but greater than 0 ft thick, bottom elevation of layer 6 is set to bottom elevation of dolomite;
- If dolomite is absent, the layer is "pinched" at this location and the bottom elevation of layer 6 is set to 0.5 ft below bottom elevation of layer 5.

Layer 7 (Silurian dolomite)

- If dolomite thickness is greater than 20 ft, bottom elevation of layer 7 is set to bottom elevation of layer 5 (i.e., of unconsolidated deposits) less dolomite thickness;
- If dolomite thickness is less than 20 ft, the layer is "pinched" at this location and the bottom elevation of layer 7 is set to 0.5 ft below bottom elevation of layer 6.

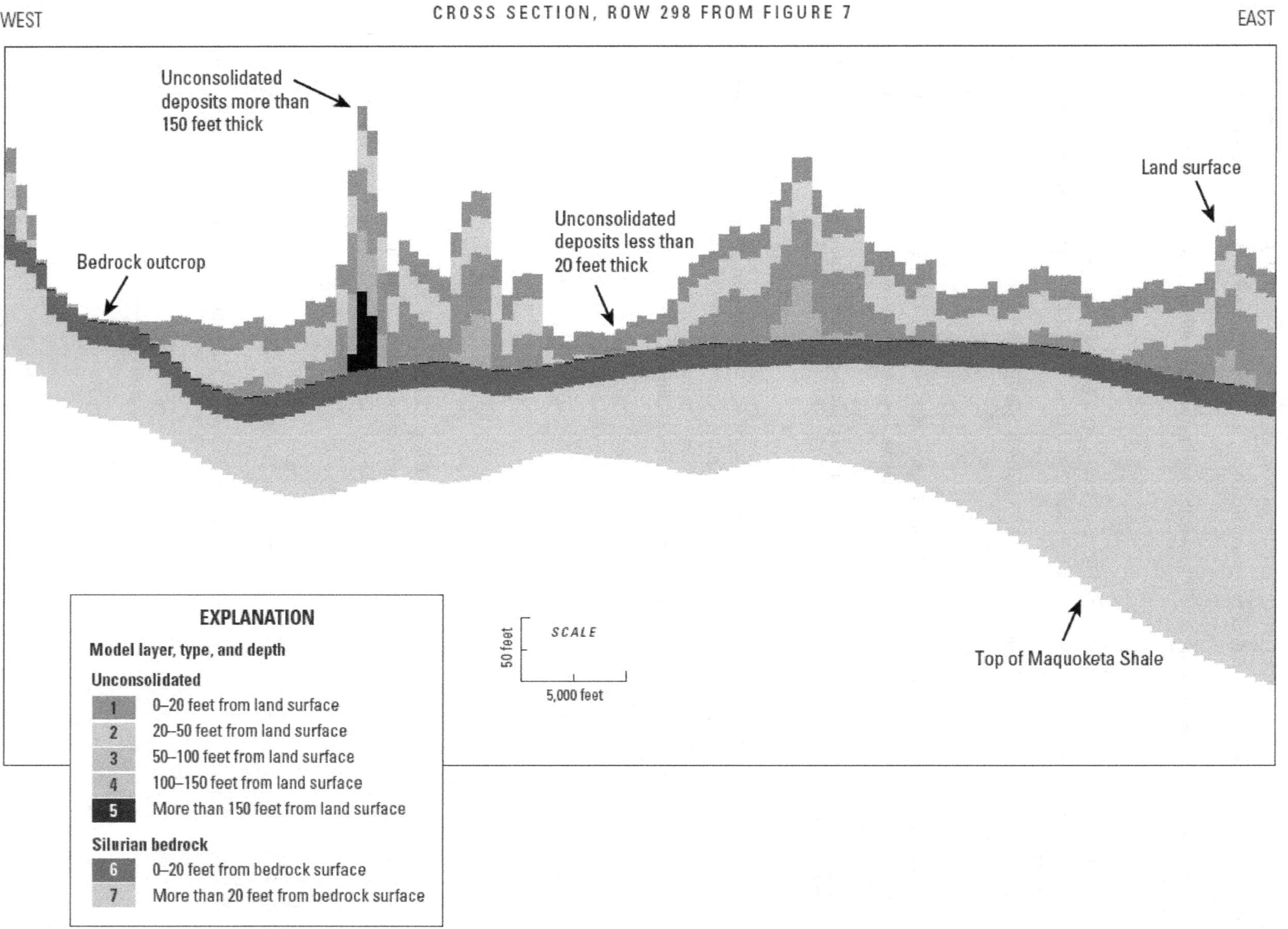

Figure 8. Model layering: example vertical section.

The elevation data from the DTM files were interpolated to 500-ft centers over the model domain. The 500-ft spacing is appropriate to the accuracy of the smoothed LIDAR data supporting the estimation of the land surface. The interpolation was performed using the "natural neighbor" algorithm (Sibson, 1981), which finds the closest subset of data to a query point and applies weights based on proportionate areas in order to interpolate a value. The scheme guarantees interpolated elevations are within the range of the data used; it does not infer trends and will not produce peaks, pits, ridges, or valleys not already represented by the elevation data. The node center locations for a single model cell were paired with the closest location of the 500-ft centers and the corresponding interpolated land- surface elevation was assigned to the cell. Note that clusters of 16 cells on 125-ft centers in the model nearfield share a single land-surface elevation from the background interpolation at 500-ft centers.

The top of layer 1 corresponds to the land surface assigned to the model cells. Almost the entire surface-water network is represented by model boundary conditions associated with layer 1 cells. The few exceptions are limited to large-area cells in the model farfield where downcutting of a stream produces a streambed that is more than 20 ft below the land-surface elevation assigned the cell. It is worth emphasizing that the use of thin layers (that is, 20-ft maximum thickness for layer 1 and 30 ft maximum thickness for layer 2) permits a relatively precise calculation of vertical gradients in the vicinity of surface-water features.

The bottom of layer 7 corresponds to the bottom of the Silurian dolomite (shown schematically in fig. 3). Over most of the model, the bottom of the Silurian dolomite is equivalent to the top of the Maquoketa shale, the regional confining unit separating the shallow and deep parts of the flow system. However, over two small areas of the model farfield (at the northwestern and southwestern corners of the grid), the Maquoketa shale subcrop boundary implies that a deeper Paleozoic unit (the Sinnipee dolomite or the St. Peter sandstone) lies at the bottom of the Upper Fox River Basin models. The bottom of the model at any row/column location corresponds not only to a stratigraphic boundary but also to an upward or downward flux boundary condition, which is described in the next subsection.

3.1.3 Flux Boundary Conditions

From the standpoint of edge boundary conditions, the Upper Fox River Basin model can be considered an inset model extracted from the SEWRPC regional model for southeastern Wisconsin. (Note that properties and boundary conditions internal to the model grid have a less direct relation to the parent model, as is explained in subsequent subsections). The Upper Fox River Basin model is entirely contained within the SEWRPC regional model, both in terms of lateral extent (the regional model extends over seven full counties in southeastern Wisconsin, including Waukesha and Washington Counties) and in terms of vertical extent (the SEWRPC model extends vertically (fig. 1*C*) from land surface to the Mount Simon sandstone at the bottom of the Cambrian-Ordovician deposits). The regional model contains 18 layers, including 2 layers representing the unconsolidated material and 3 layers representing the Silurian dolomite. The top layer of the regional model is up to 100 ft thick (Feinstein and others, 2005a).

As part of a recently-completed water-supply study for southeastern Wisconsin (Southeastern Wisconsin Regional Planning Commission, 2010), the original SEWRPC model was updated with 2005 pumping estimates, and the transient simulation extended beyond the original endpoint of the model in 2000. The output from the regional model corresponding to 2005 pumping rates is extracted as input to the Upper Fox River Basin model in the form of edge boundary fluxes. To facilitate the transfer of the lateral fluxes, the edge rows and columns of the inset Upper Fox models have been assigned the same widths as the cells in the parent regional model—2,500 ft per side. Layers in the regional model are associated with layers in the inset model on the basis of thickness and stratigraphy. Node centers in the regional model also are paired with boundary nodes of the inset model—typically multiple nodes in the Upper Fox models are associated with a single regional node from the SEWRPC model. The fluxes are inserted by means of the WEL package in MODFLOW at the boundary cells of the Upper Fox models.

The regional model contributes inflows and outflows at different layer depth intervals with respect to the inset area. When flows are tabulated, unconsolidated and dolomite layers along each side of the inset model both lose and gain groundwater from the surrounding region (table 2). However, on balance the net flux derived from the regional SEWRPC model is outward across all four sides of the Upper Fox River Basin model domain, especially across the east edge of the model where both the topography and the dip of the bedrock tends to slope eastward.

The bottom flux boundary condition for the inset model also is derived from the SEWRPC regional model. The fluxes from the regional model are available at a 2,500-ft spacing and, therefore, a single regional flux value is assigned over an area represented by multiple bottom Upper Fox model cells (equal to 16 cells for the nearfield). Again, the WEL package is used to insert the flux and the direction of flow varies by location. On balance the vertical flow is outward from the bottom of the Upper Fox models (table 2), which is in line with the regional tendency of water to leak downward from the shallow to the deep aquifer system as a result of large pumping centers drawing from the Cambrian-Ordovician units in Cities such as Waukesha.

The net flux out of the sides and bottom of the shallow aquifer system incorporated in the Upper Fox model domain is on the order of 7 million gallons per day (Mgal/d). This net total is about 10 percent of the total recharge to the top of the groundwater system discussed below.

Table 2. Fixed fluxes assigned to model at grid boundaries for groundwater-flow model of Upper Fox River Basin, southeastern Wisconsin.

[Flux unit is million gallons per day; unconsolidated deposits correspond to model layers 1–5; Silurian dolomite corresponds to layers 6 and 7; all layers correspond to layers 1–7]

Lateral model boundaries	Layers	Flux in	Flux out	Net flux out
North	Unconsolidated deposits	0.299	0.313	0.014
	Silurian dolomite	.153	.189	.036
	All layers	.452	.502	.050
East	Unconsolidated deposits	.058	.470	.412
	Silurian dolomite	.054	1.281	1.227
	All layers	.113	1.751	1.638
South	Unconsolidated deposits	.090	1.068	.978
	Silurian dolomite	.051	.333	.282
	All layers	.141	1.400	1.259
West	Unconsolidated deposits	.493	1.719	1.226
	Silurian dolomite	.108	.113	.005
	All layers	.601	1.832	1.231
All sides	Unconsolidated deposits	.940	3.570	2.630
	Silurian dolomite	.366	1.916	1.550
	All layers	1.307	5.485	4.178
Bottom model boundary		.31	3.48	3.17
All model boundaries		1.62	8.96	7.35

3.2 Hydraulic Conductivity Zones

The focus of the Upper Fox River Basin model is the most shallow part of the saturated flow system where groundwater/surface-water interactions occur. The water table in the model domain is almost always within the heterogeneous unconsolidated deposits and, therefore, special attention has been given to characterizing their lateral and vertical hydraulic conductivity . The assignment of hydraulic conductivity zones to the underlying dolomite, by contrast, is relatively simple and is inherited from the parent SEWRPC model.

3.2.1 Unconsolidated Deposits

The assignment of hydraulic conductivity to unconsolidated deposits in this study is based on textural interpretations of the subsurface derived from a database consisting mostly of water-well driller logs compiled from three sources:

- digital records of well construction logs compiled by the Wisconsin Department of Natural Resources (2009) for the period from 1988 through December 2008,
- microfiche records at the University of Wisconsin-Milwaukee library for the period before 1988, and
- records compiled by the USGS as part of the construction of the Lake Michigan Basin model (Arihood, 2009).

These records are supplemented by logs prepared by geologists at the Wisconsin Geological and Natural History Survey (2004), and test borings obtained from several consulting reports (GeEx, 1989; Aquifer Science and Technology, 2004, 2006, and 2010). The total database consists of about 7,000 logs (fig. 9). In the farfield of the model, the logs are located to the nearest quarter-quarter section. In the nearfield of the model, street addresses in combination with an online geographic database for Waukesha County (Waukesha County Internet Mapping site,2009–10) were used to achieve more precise locations. The database locations are provided in latitude/longitude and in Wisconsin Transverse Mercator North American Datum of 1983 (NAD 83) projection.

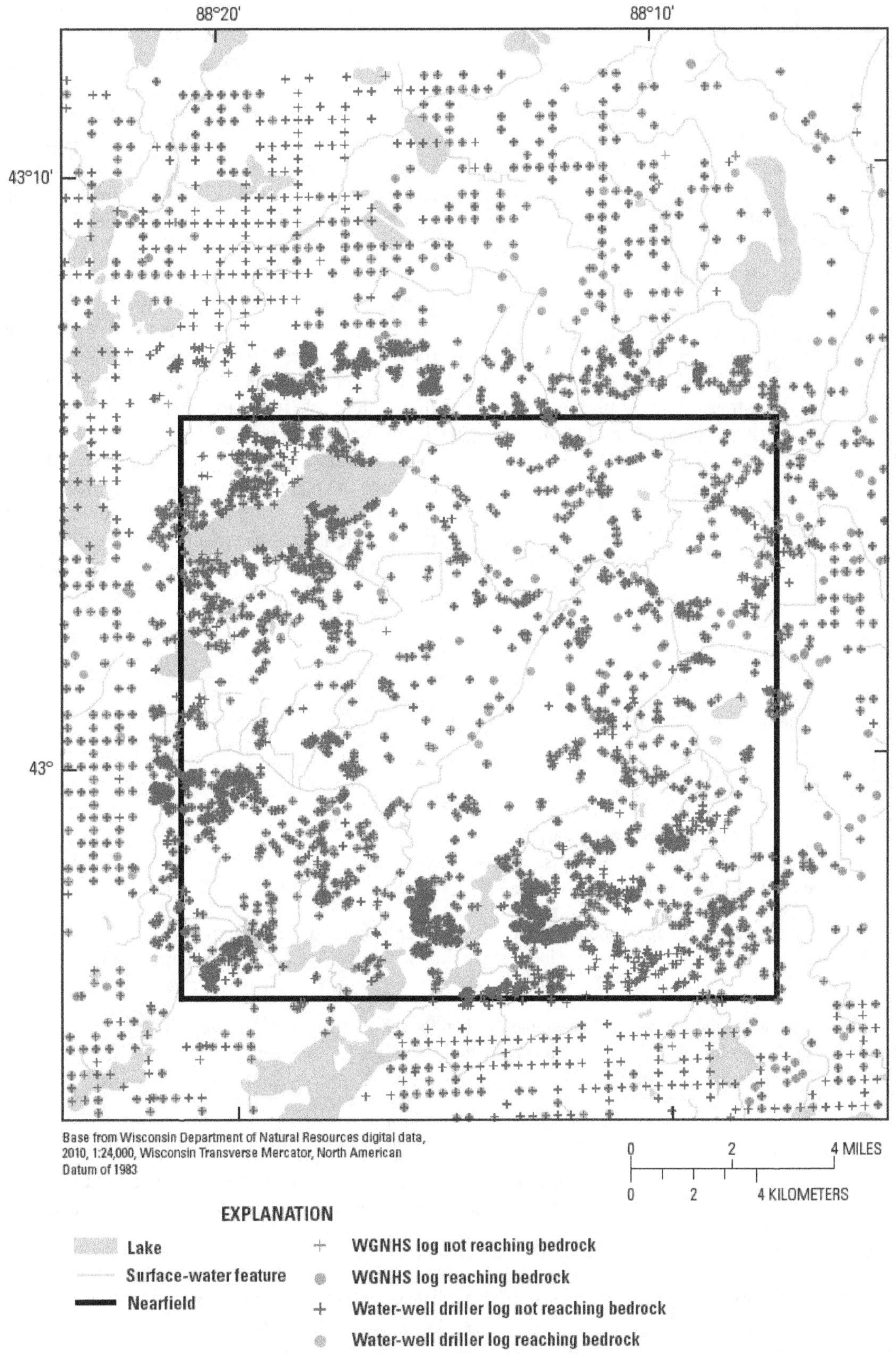

Figure 9. Well-log database. (Well-log data interpreted from Wisconsin Department of Natural Resources, 2009, and Wisconsin Geological and Natural History Survey, 2004, digital-data sources.) (WGNHS, Wisconsin Geological and Natural History Survey.)

Most of the logs penetrate the entire unconsolidated sequence and extend into the bedrock (fig. 9). All logs provide information as a function of depth. For the model database, depths were converted to absolute elevations using the log location and a 2-ft contour land-surface coverage provided by the Waukesha County Internet Mapping site.

The logs provide descriptions of the texture of the unconsolidated deposits as a function of depth intervals. Water-well drillers use many terms to characterize the deposits. The initial dataset contained 357 unique identifiers, most applied to unconsolidated deposits. These identifiers were grouped in three classes:

- descriptions limited to fine-grained textures (clay, silt, silty clay, hardpan, and others)
- descriptions combining fine- and coarse-grained textures (silty sand, gravelly clay, silt and sand, and others)
- descriptions limited to coarse-grained textures (sand, gravel, sand and gravel, and others)

Each depth interval in a log corresponding to a single description was assigned to one of the three texture classes. However, in general, the depth intervals recorded in the logs do not match the depth intervals corresponding to the model layering scheme. Typically, a layer depth interval contains more than one texture class (fig. 10). In order to characterize the layer depth intervals over the vertical extent of each log, a set of rules was adopted that depend on the dominant texture reported within the model layer depth interval. At a log location, the depth interval for a layer is considered *dominantly fine (facies 1)* if 90 percent of the layer thickness contains descriptions reported as fine texture, *dominantly coarse (facies 5)* if 90 percent of the layer thickness is reported as coarse texture, and *mixed fine and coarse (facies 3)* if 75 percent of the layer thickness is reported as mixed facies. In addition, two transitional facies are applied if none of the above criteria is met. If the layer depth interval contains more of texture class 1 and 2 than texture class 2 and 3, it is assigned the *relatively fine (facies 2)*. Otherwise, it is assigned the *relatively coarse (facies 4)*. The logic for the five possible facies assignments to model- layer intervals at log locations is systematically presented in table 3. An example assignment to facies 4 for a depth interval corresponding to model layer 3 is illustrated in figure 10.

Table 3. Logic for assignment of unconsolidated facies in model layers 1–5.

[ft, foot; %, percent]

- All driller log intervals are assigned to one of three texture classes:

  Class 1 = All fines (clay, silt, silty clay, clayey silt)
  Class 2 = Mixed fines and coarse (silt with sand, sandy clay, clayey sand, silt and gravel)
  Class 3 = Sand and coarser (sand, gravel, sand and gravel)

- Within each model layer interval (0- to 20-ft depth, 20- to 50-ft depth, 50- to 100-ft depth, 100- to 150-ft depth, more than 150-ft depth), the three texture classes at driller log locations are converted into five facies according to the following rules:

Step 1:

If at least 90% of the layer interval corresponds to class 1, then the facies is "dominantly fine"= facies 1
If at least 75% of the layer interval corresponds to class 2, then the facies is "dominantly mixed" = facies 3
If at least 90% of the layer interval corresponds to class 3, then the facies is "dominantly coarse" = facies 5

Step 2:

If none of the criteria are meet in Step 1, then
If more than 50% of the layer interval consists of class 1 and class 2, then the facies is "relatively fine" = facies 2
If more than 50% of the layer interval consists of class 2 and class 3, then the facies is "relatively coarse" = facies 4

See figure 10 for visualization.

- The facies assignments at driller log locations are interpolated to node centers on a 500-ft by 500-ft grid for each model layer.

  Two methods are used for interpolation—one is fine-favored and one is coarse-favored (fig. 12). The 500-ft by 500-ft grid assignments are transferred to the model grid based by associating the model row/column node centers with the nearest node center on the 500 ft by 500 ft grid. The distribution of facies in the fine-favored and coarse-favored models is shown in figures 14 and 15.

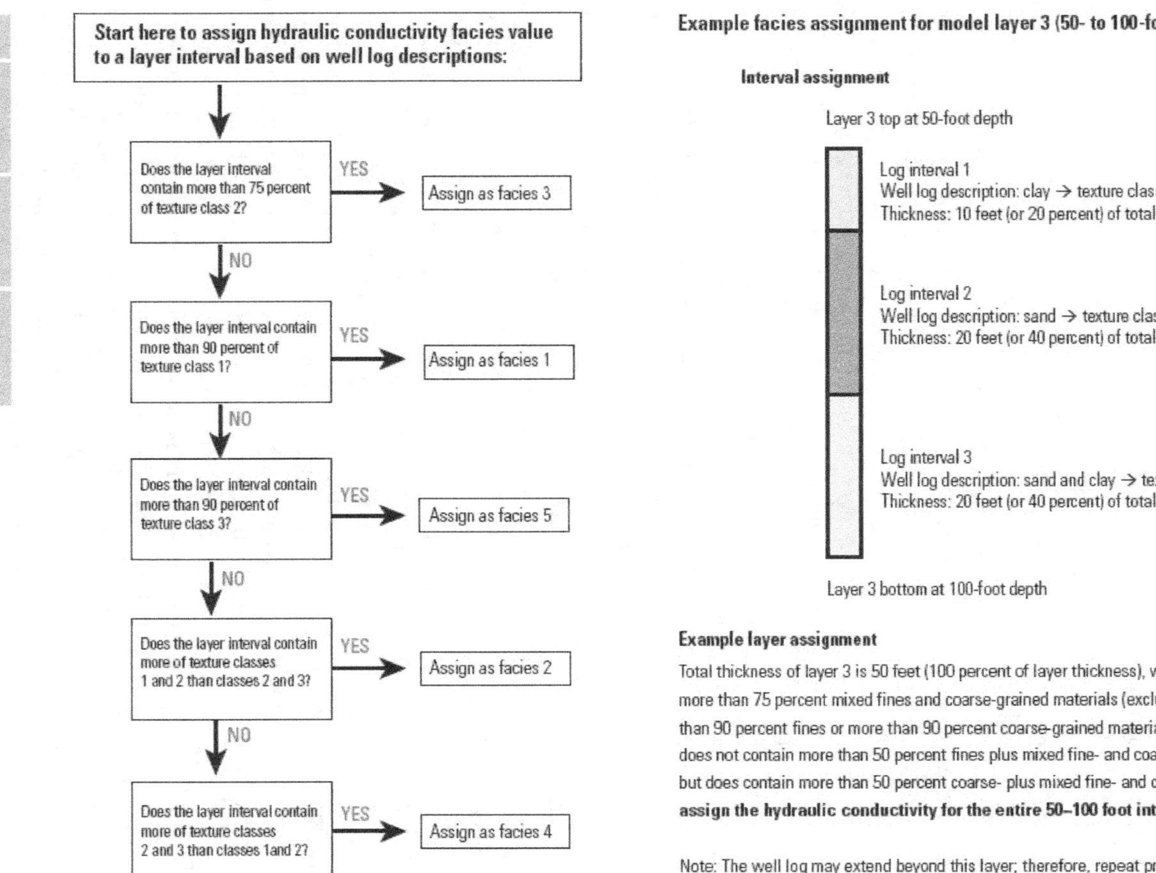

Hydraulic conductivity facies explanation

Value	Assignment	Lithologic description	Examples of corresponding log descriptions
1	dominantly fine	more than 90 percent fine	clay, silt, clayey till, silty till
2	relatively fine	more fine than coarse	sandy clay and silt
3	mixed fines and coarse	more than 75 percent mixed fines and coarse	loamy till, sand and clay
4	relatively coarse	more coarse than fine	silty sand and gravel
5	dominantly coarse	more than 90 percent coarse	sand, gravel, sand and gravel

Figure 10 Logic for assigning facies to layer intervals

Maps of the distribution of facies at log locations for each of the unconsolidated layer depth intervals (figs. 11*A–E*) show fine-grained deposits are more common in the eastern part of the model domain (Oak Creek Formation) and coarse-grained deposits are more common in the western part (Horicon Member). The plots also show the active and pinched locations of the five unconsolidated layers. Note that unconsolidated layer 5, corresponding to a depth of more than 150 ft, is almost completely pinched inside the model nearfield (fig. 11*E*).

It is helpful to visualize the textural trends across the model domain not only by plotting the distribution of log facies but also by interpolating the percent of the dominantly coarse facies from the log locations over the domain area. In this way, it is evident where the unconsolidated sediments are likely to be most permeable and most resistive to flow because of high and low sand content, respectively. The interpolation results (shown for model layer 1 corresponding to 0–20 ft depth in fig. 12*A* and for model layer 3 corresponding to 50–100 ft depth in fig. 12*B*) reveal that in general about one-third of the sediments correspond to sand or gravel texture, but that the eastern side of the model domain is poor in coarse sediments, the western side is rich in coarse sediments, and the central region displays a complicated, interfingered pattern. The three texture zones match the distribution of the glacial units discussed in the "Conceptual Model" section—the clayey Oak Creek Formation to the east, the sandy Horicon Member to the west, and mixed New Berlin Member in the middle of the study area.

3.2.1.1 Fine-Favored and Coarse-Favored Zonation

The five facies descriptions applied to logs by layer-depth interval are the basis for the mapping of five unconsolidated hydraulic conductivity (K) zones across all the active cells in each layer. However, the necessity to interpolate facies from model cells where logs are present to neighboring cells where no logs are present adds a crucial element of uncertainty to the mapping. In order to explicitly account for the uncertainty, an interpolation method is applied which, as explained in the "Conceptual Model" section, allows the continuity of fine-grained facies (facies 1 and 2) to be favored or the continuity of coarse-grained facies (facies 4 and 5) to be favored. The method uses a nearest neighbor approach in the following way: (a) if a cell in given layer-depth interval contains one or more logs within its area, the facies from those logs are used to assign a K zone to the cell, such that the finest facies encountered in the cell is used for the fine-favored mapping and the coarsest facies encountered for the coarse-favored mapping; (b) if a cell does not contain a log in a layer-depth interval, circles of increasing radii centered at the cell center are superimposed on the grid until the circle contains one or more logs, such that the finest facies encountered in the circle is used for the fine-favored mapping and the coarsest facies encountered for the coarse-favored mapping. The first search circle is assigned a radius of 708 ft (equal to the diagonal length of a cell 500 ft per side); if no log is encountered, the search circle is widened by a factor of two to 1,416 ft; if no log is encountered, the search circle is widened to 2,124 ft, and so on.

The method for interpolating the facies, that is K zones, into cells without logs is shown graphically in figure 13. The results of the procedure are shown for an example cross section in figure 14 and in plain view for each model layer in figure 15. In each case, the fine-favored interpretation is contrasted with the coarse-favored.

Note that in model layer 5 (unconsolidated material more than 150 ft below land surface, associated with bedrock valleys), the configuration of K zones is identical for the fine-favored and coarse-favored models (fig. 15*E*). The zones do not correspond to interpolation from logs (of which there are very few at this depth), but rather to a broad zonation of coarse material in the bedrock valley associated with the Bark River in the northwestern part of the model domain and mixed material elsewhere, including the Troy Bedrock Valley, which cuts through the southern part of the model domain. The evidence for this pattern is derived from previous studies, which treated the sediment in the bedrock valleys (Batten and Conlon, 1993; Feinstein and others, 2011). The simple zonation of the K distribution in layer 5 is almost entirely limited to the farfield of the model domain because layer 5 is pinched out in most of the nearfield.

Neither the coarse-favored or the fine-favored version is correct in the sense of being a true representation of the subsurface at a given depth interval but are useful in the different way connectivity is treated between high K zones. For example, the small box near the south end of the model nearfield in the plots in figure 15 shows the different proportion of fine and coarse sediments along one riparian zone of the Fox River generated by the two interpretations. These contrasting patterns have implications for the ability of a well located near the river to induce flow from the river. In the example cross sections in figure 14, it is evident that vertical flow is facilitated in the coarse-favored case relative to the fine-favored case because of the greater lateral and vertical connectivity of the coarse-dominated hydraulic conductivity zone (colored red in the plots). This outcome is important because the vertical hydraulic conductivity (K_v) of the dominantly coarse K zone is expected to be elevated with respect to the other four zones given the near or total absence of fine-grained material in facies 5 (dominantly coarse). As a result, flow, for example, induced from a stream toward a nearby well screened 50 or 100 ft below land surface would likely be transmitted more easily in the coarse-favored case because there are more areas where there is vertical continuity of the dominantly coarse facies.

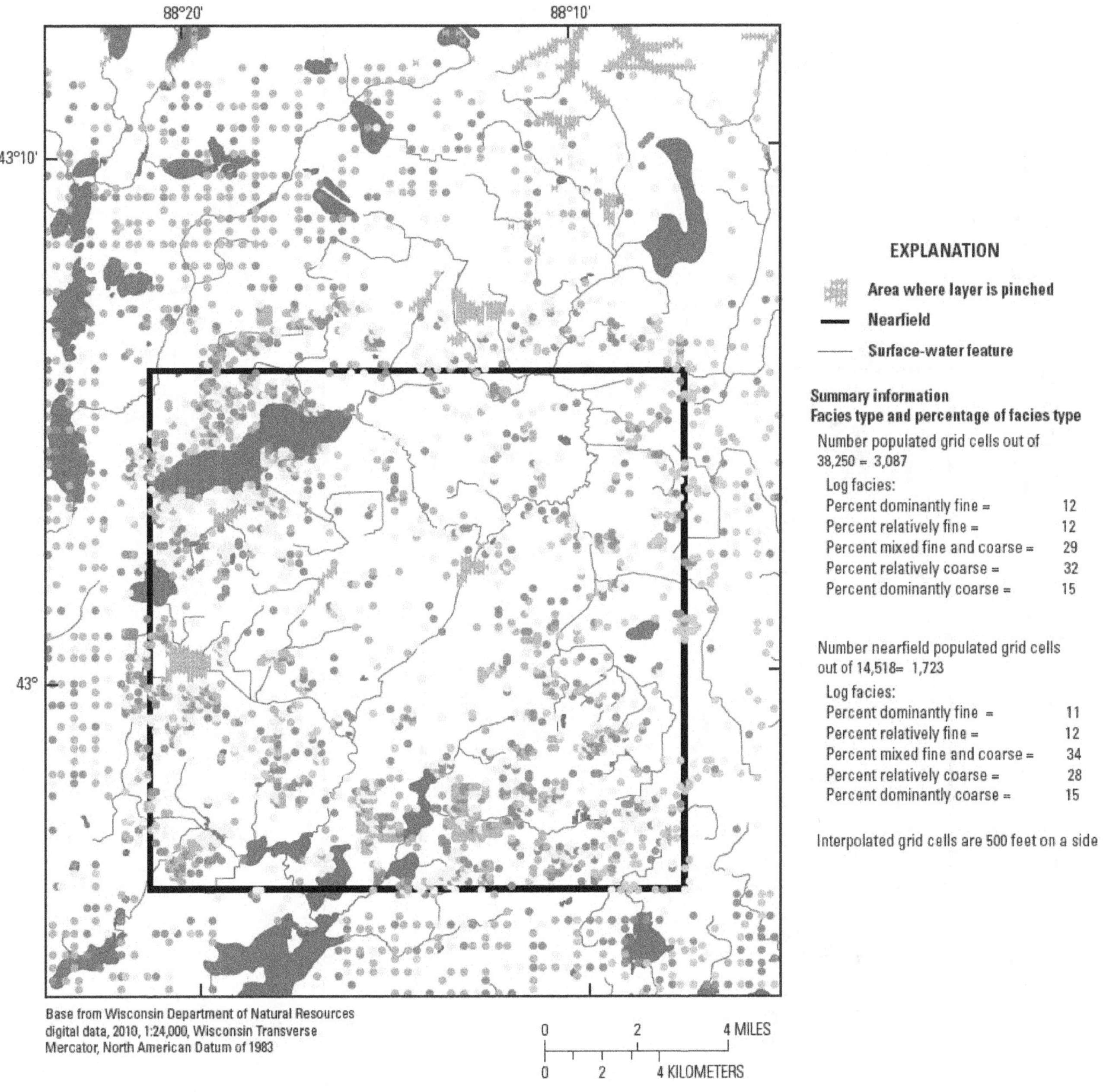

Symbol	Code	Assignment	Lithologic description
	1	Dominantly fine	Greater than 90 percent fine
	2	Relatively fine	More fine than coarse
	3	Mixed	Greater than 75 percent mixed fines and coarse
	4	Relatively coarse	More coarse than fine
	5	Dominantly coarse	Greater than 90 percent coarse

Figure 11*A*. Classification of log descriptions by dominant facies—layer 1 (0–20 foot depth). (Well-log data interpreted from Wisconsin Department of Natural Resources, 2009, and Wisconsin Geological and Natural History Survey, 2004, digital-data sources.)

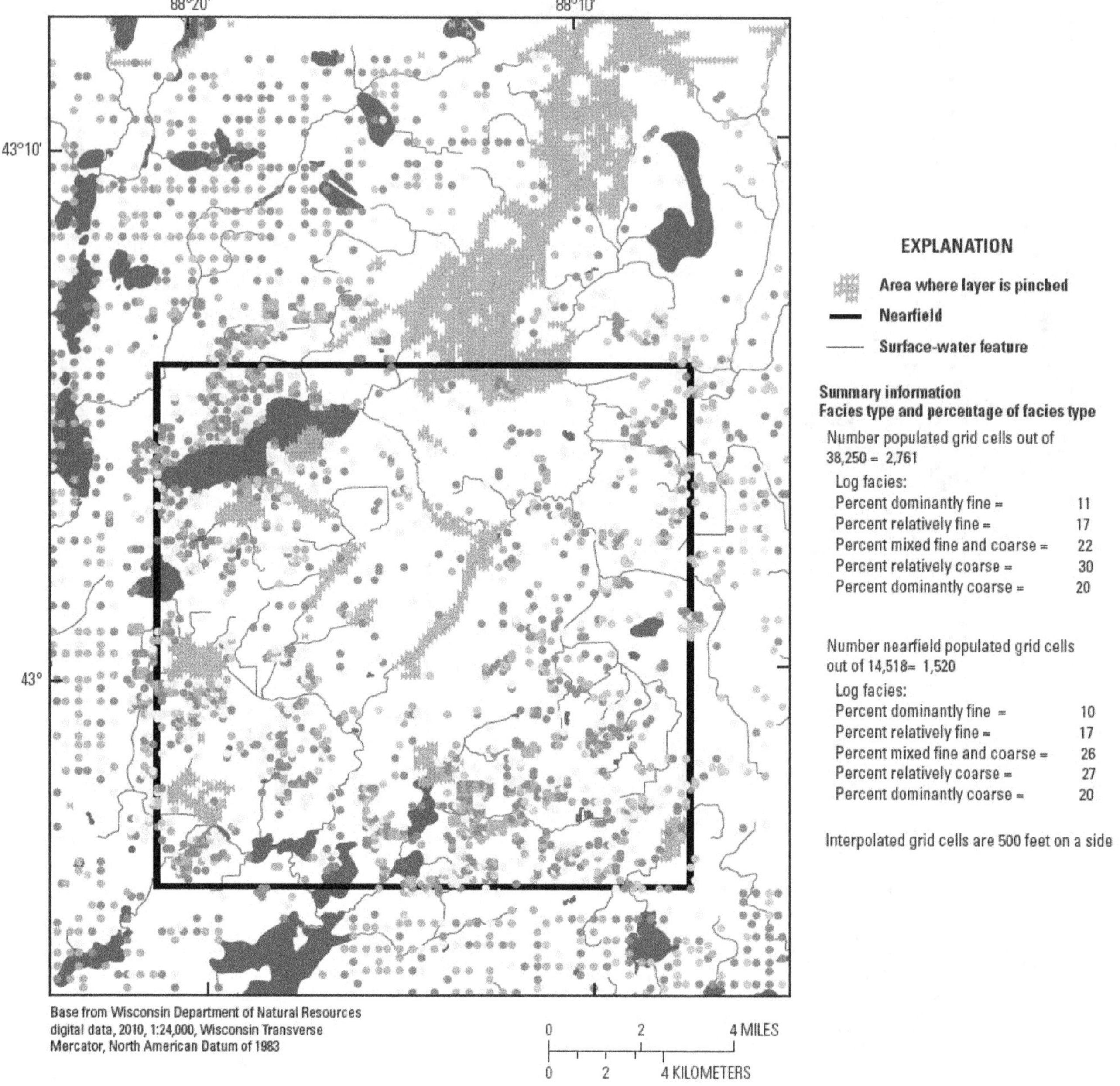

Symbol	Code	Assignment	Lithologic description
	1	Dominantly fine	Greater than 90 percent fine
	2	Relatively fine	More fine than coarse
	3	Mixed	Greater than 75 percent mixed fines and coarse
	4	Relatively coarse	More coarse than fine
	5	Dominantly coarse	Greater than 90 percent coarse

Figure 11*B*. Classification of log descriptions by dominant facies—layer 2 (20–50 foot depth). (Well-log data interpreted from Wisconsin Department of Natural Resources, 2009, and Wisconsin Geological and Natural History Survey, 2004, digital-data sources.)

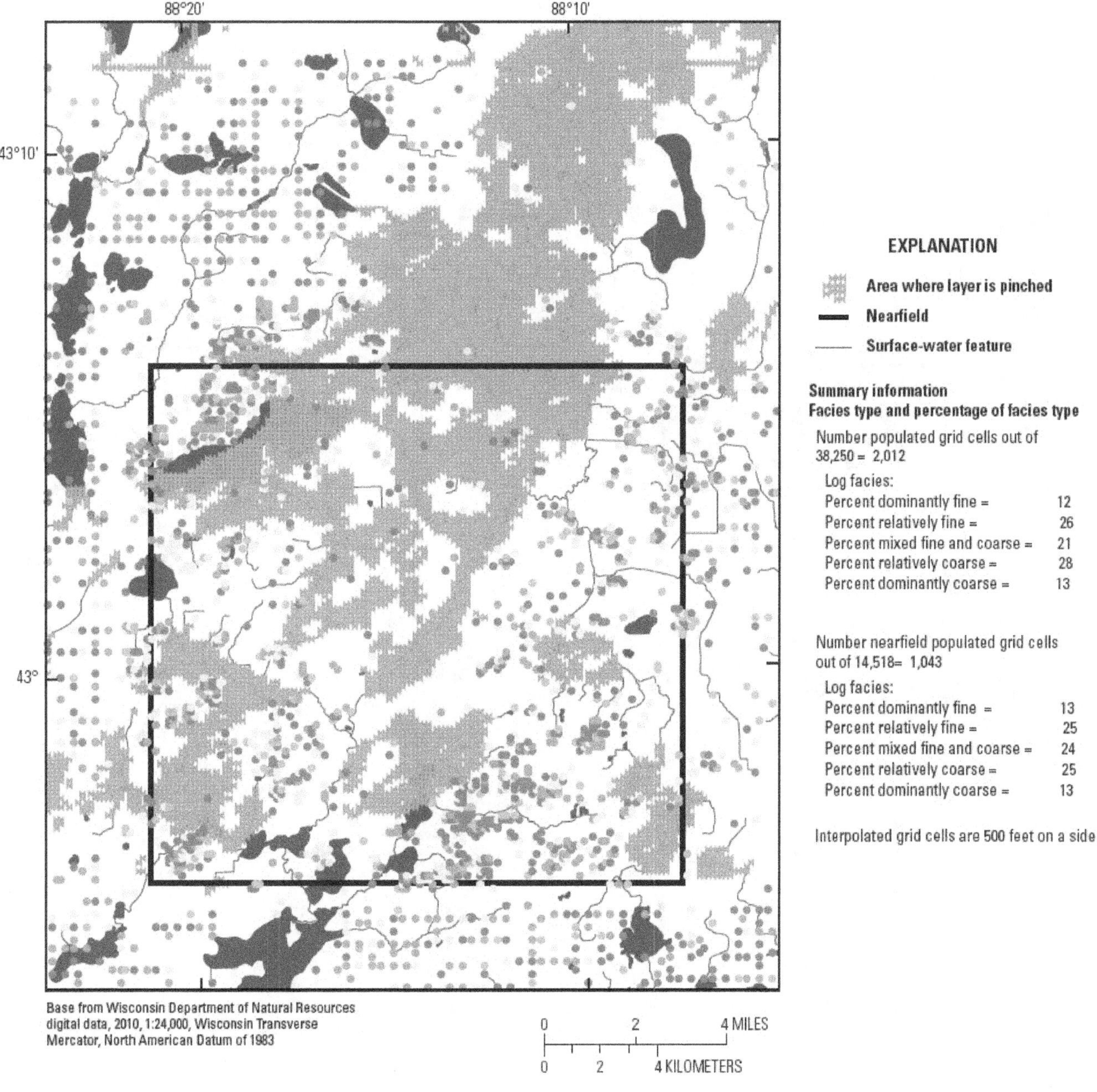

Symbol	Code	Assignment	Lithologic description
	1	Dominantly fine	Greater than 90 percent fine
	2	Relatively fine	More fine than coarse
	3	Mixed	Greater than 75 percent mixed fines and coarse
	4	Relatively coarse	More coarse than fine
	5	Dominantly coarse	Greater than 90 percent coarse

Figure 11*C*. Classification of log descriptions by dominant facies—layer 3 (50–100 foot depth). (Well-log data interpreted from Wisconsin Department of Natural Resources, 2009, and Wisconsin Geological and Natural History Survey, 2004, digital-data sources.)

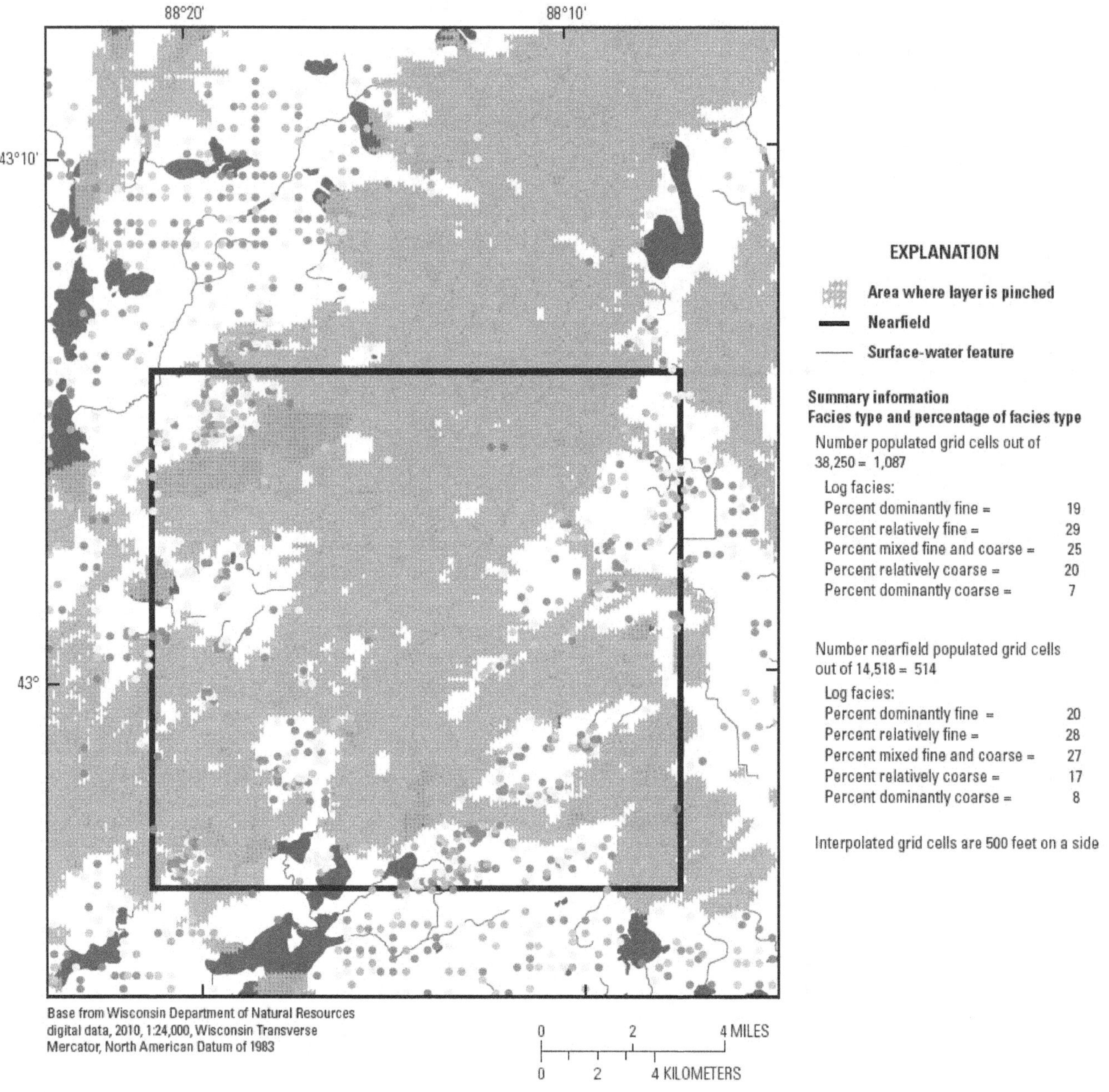

Symbol	Code	Assignment	Lithologic description
	1	Dominantly fine	Greater than 90 percent fine
	2	Relatively fine	More fine than coarse
	3	Mixed	Greater than 75 percent mixed fines and coarse
	4	Relatively coarse	More coarse than fine
	5	Dominantly coarse	Greater than 90 percent coarse

Figure 11*D*. Classification of log descriptions by dominant facies—layer 4 (100–150 foot depth). (Well-log data interpreted from Wisconsin Department of Natural Resources, 2009, and Wisconsin Geological and Natural History Survey, 2004, digital-data sources.)

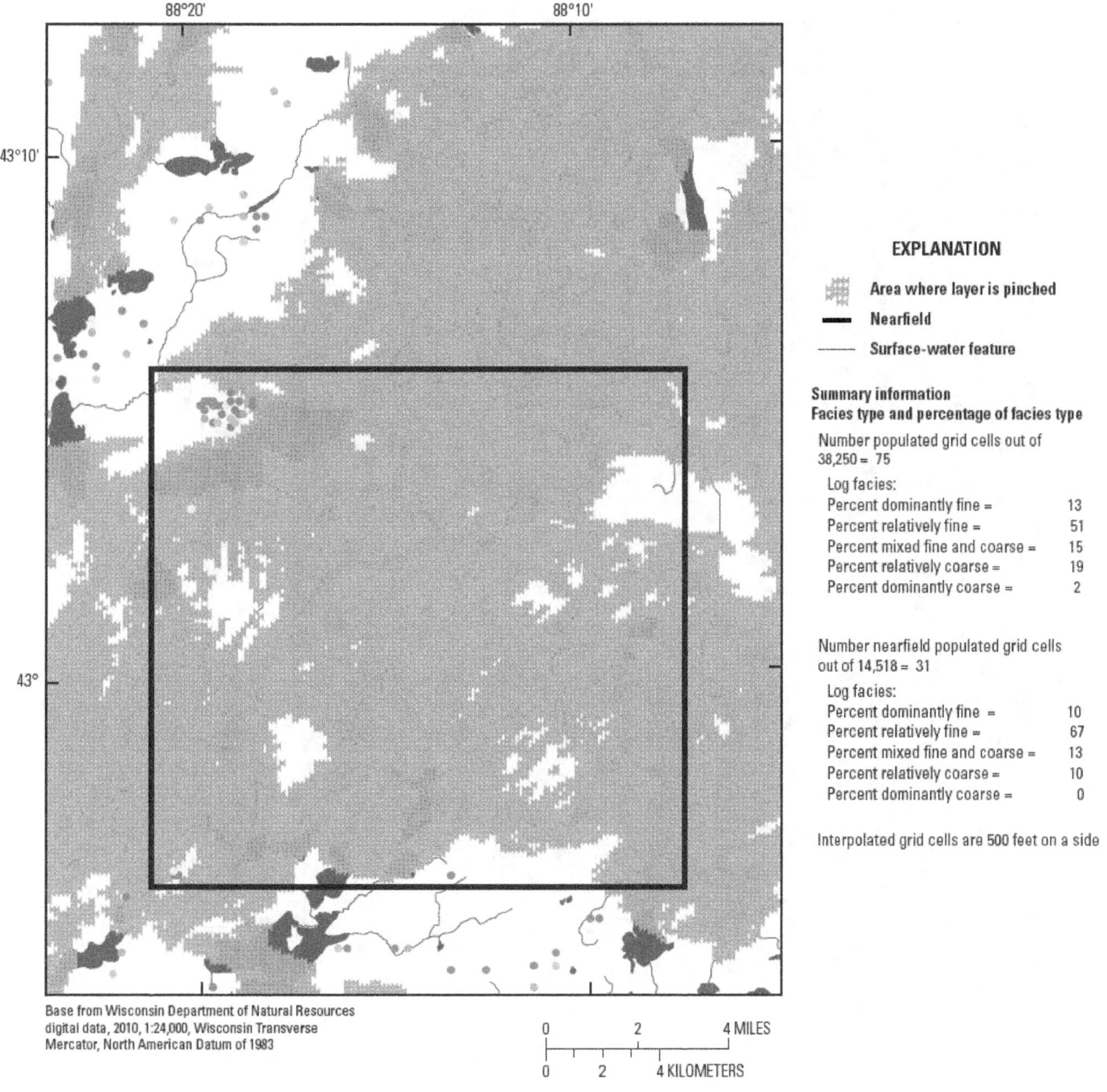

Symbol	Code	Assignment	Lithologic description
	1	Dominantly fine	Greater than 90 percent fine
	2	Relatively fine	More fine than coarse
	3	Mixed	Greater than 75 percent mixed fines and coarse
	4	Relatively coarse	More coarse than fine
	5	Dominantly coarse	Greater than 90 percent coarse

Figure 11*E*. Classification of log descriptions by dominant facies—layer 5 (more than 150 foot depth). (Well-log data interpreted from Wisconsin Department of Natural Resources, 2009, and Wisconsin Geological and Natural History Survey, 2004, digital-data sources.)

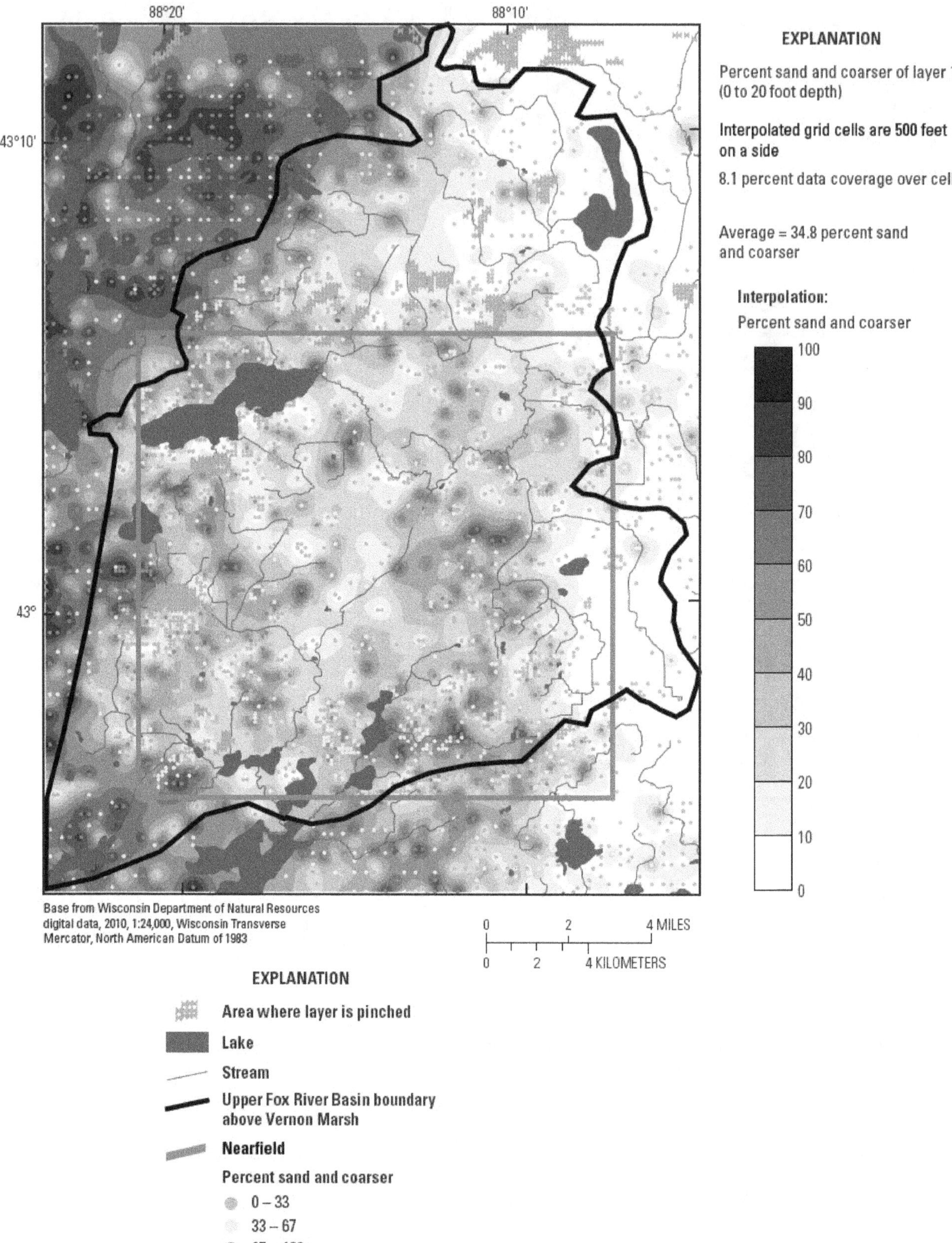

Figure 12A. Distribution of coarse deposits based on well logs for example layer intervals—layer 1 (0–20 foot depth). (Well-log data nterpreted f om Wisconsin Department of Natural Resources, 2009, and Wisconsin Geological and Natural History Survey, 2004, digital-data sources.)

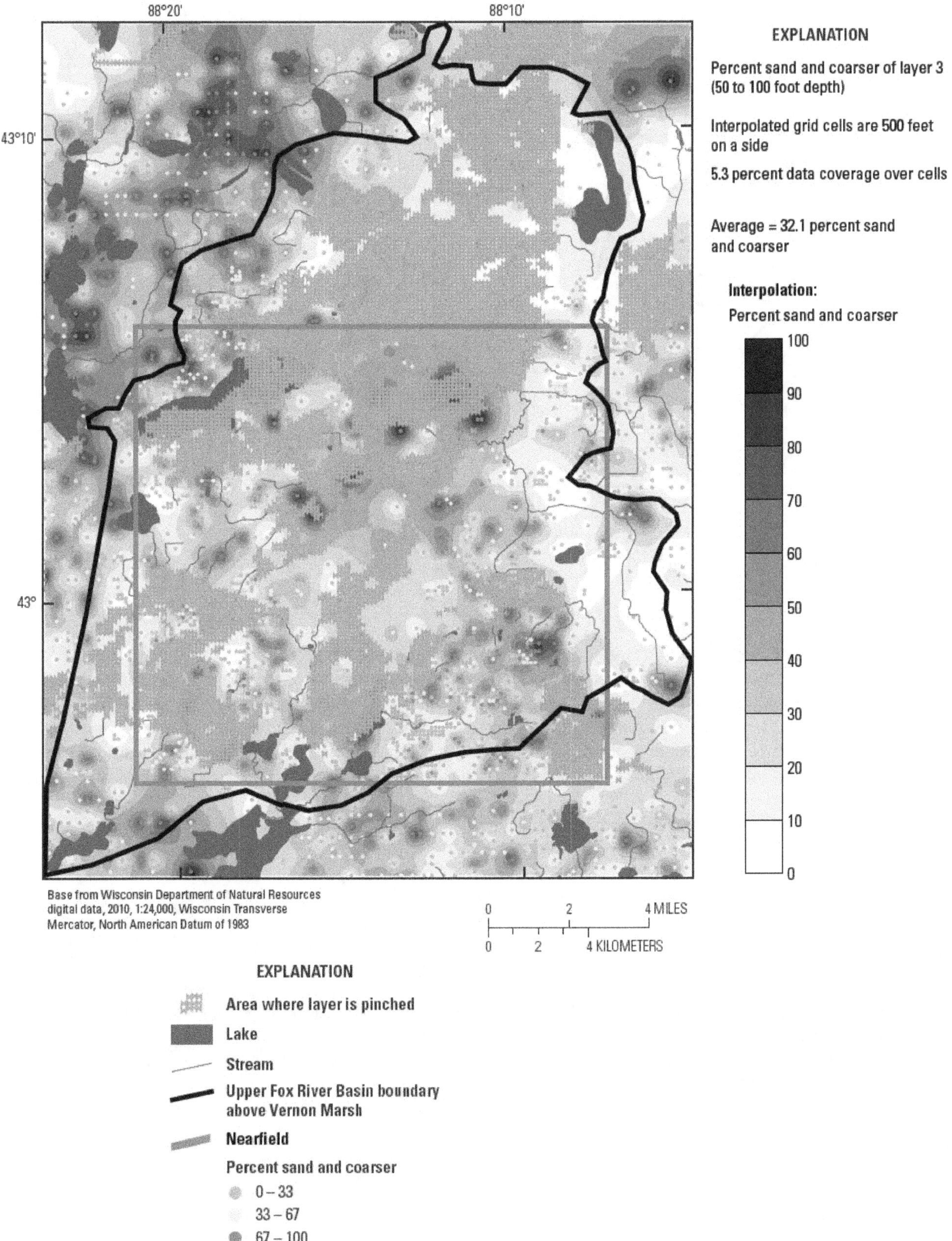

Figure 12*B*. Distribution of coarse deposits based on well logs for example layer intervals—layer 3 (50–100 foot depth). (Well-log data interpreted from Wisconsin Department of Natural Resources, 2009, and Wisconsin Geological and Natural History Survey, 2004, digital-data sources.)

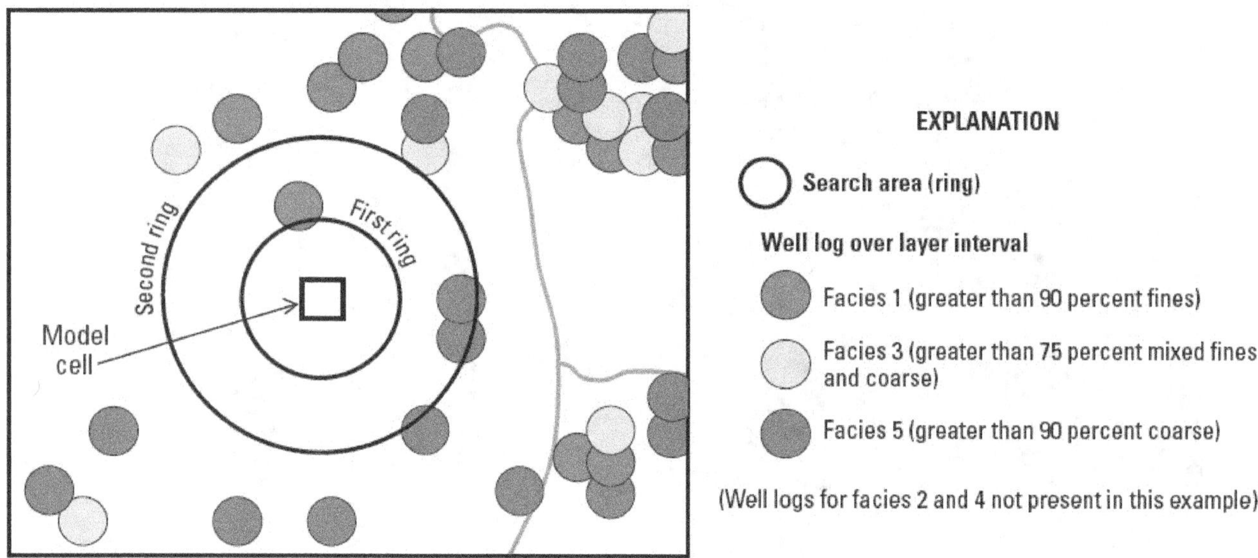

Logic of the interpolation algorithm

Starting at a model cell, if there are:

1. One or more well logs located in the first ring around model cell in question, stop at first ring and assign hydraulic conductivity facies based on the log with finest facies for fine-favored model and the log with coarsest facies for coarse-favored model. Otherwise,

2. Go to second ring. Assign hydraulic conductivity facies based on one or more well logs within the second ring using the following model logic: In the fine-favored logic model, assign the hydraulic conductivity of the model cell to the finest facies present within this second ring. In the coarse-favored logic model, assign the hydraulic conductivity of the model cell to the coarsest facies present within this second ring.

3. If there had been no logs in the first or second ring, the search area would be expanded until an annular ring around the model cell contains at least one log.

Example of application

No logs are located in first ring around the cell in question, therefore go to second ring.
Three logs are located in the second ring, therefore stop at second ring.
For the fine-favored model, assign the hydraulic conductivity of the model to value corresponding to the finest facies encountered in the second ring: in this example to the value of K1.
For the coarse-favored model, assign the hydraulic conductivity of the model to the value corresponding to the coarsest facies encountered in the second ring: in this example to the value of K5.

Figure 13. Interpolation method for fine- and coarse-favored models.

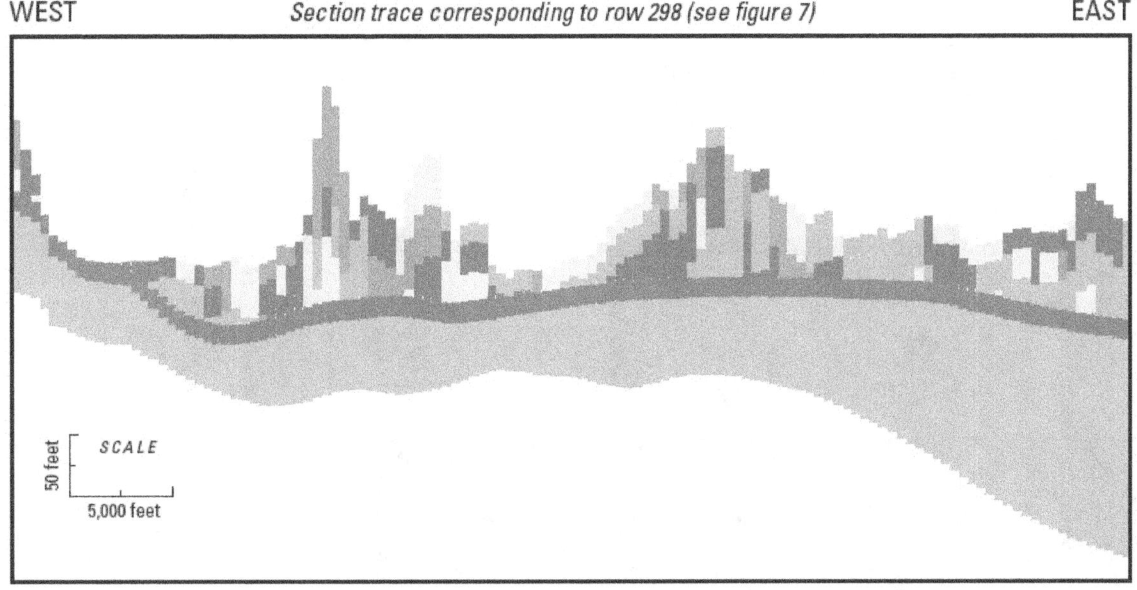

Figure 14A. Fine- and coarse-favored hydraulic conductivity zones in example vertical section—fine-favored zones.

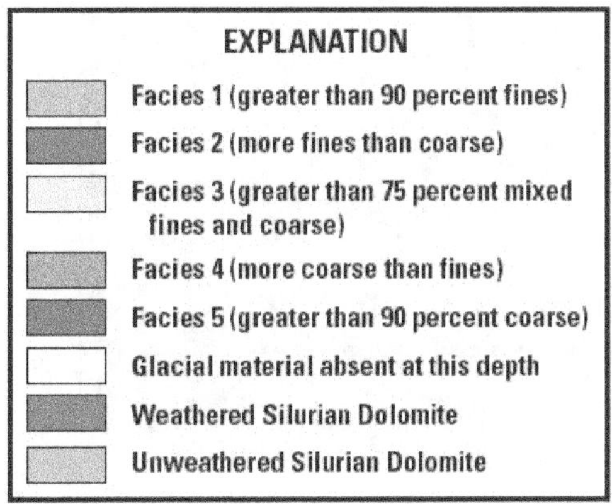

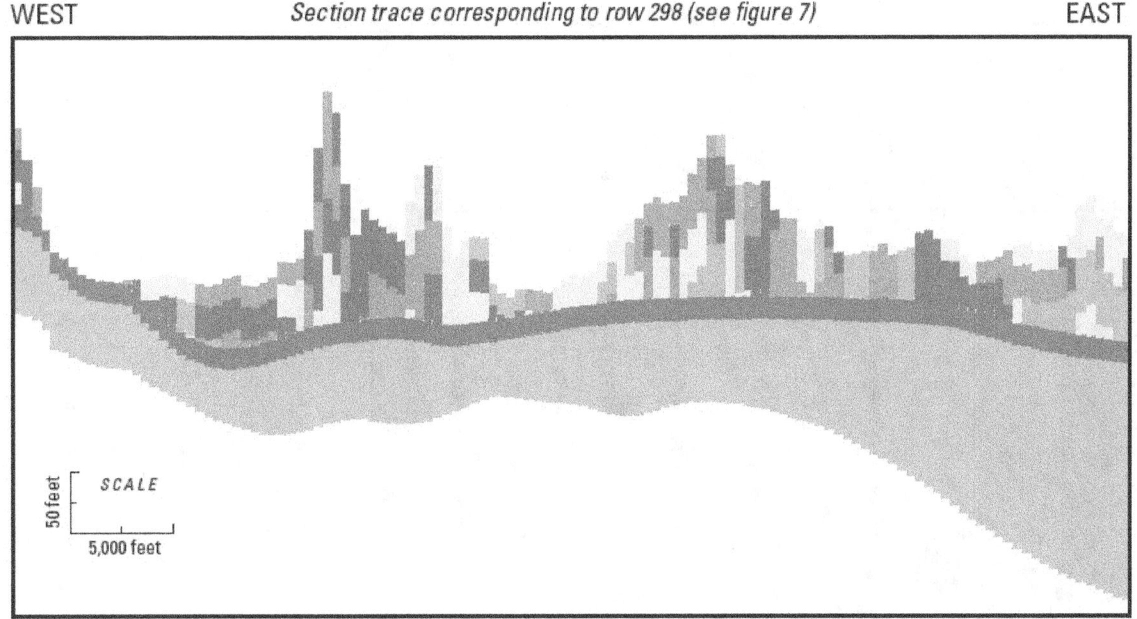

Figure 14B. Fine- and coarse-favored hydraulic conductivity zones in example vertical section—coarse-favored zones.

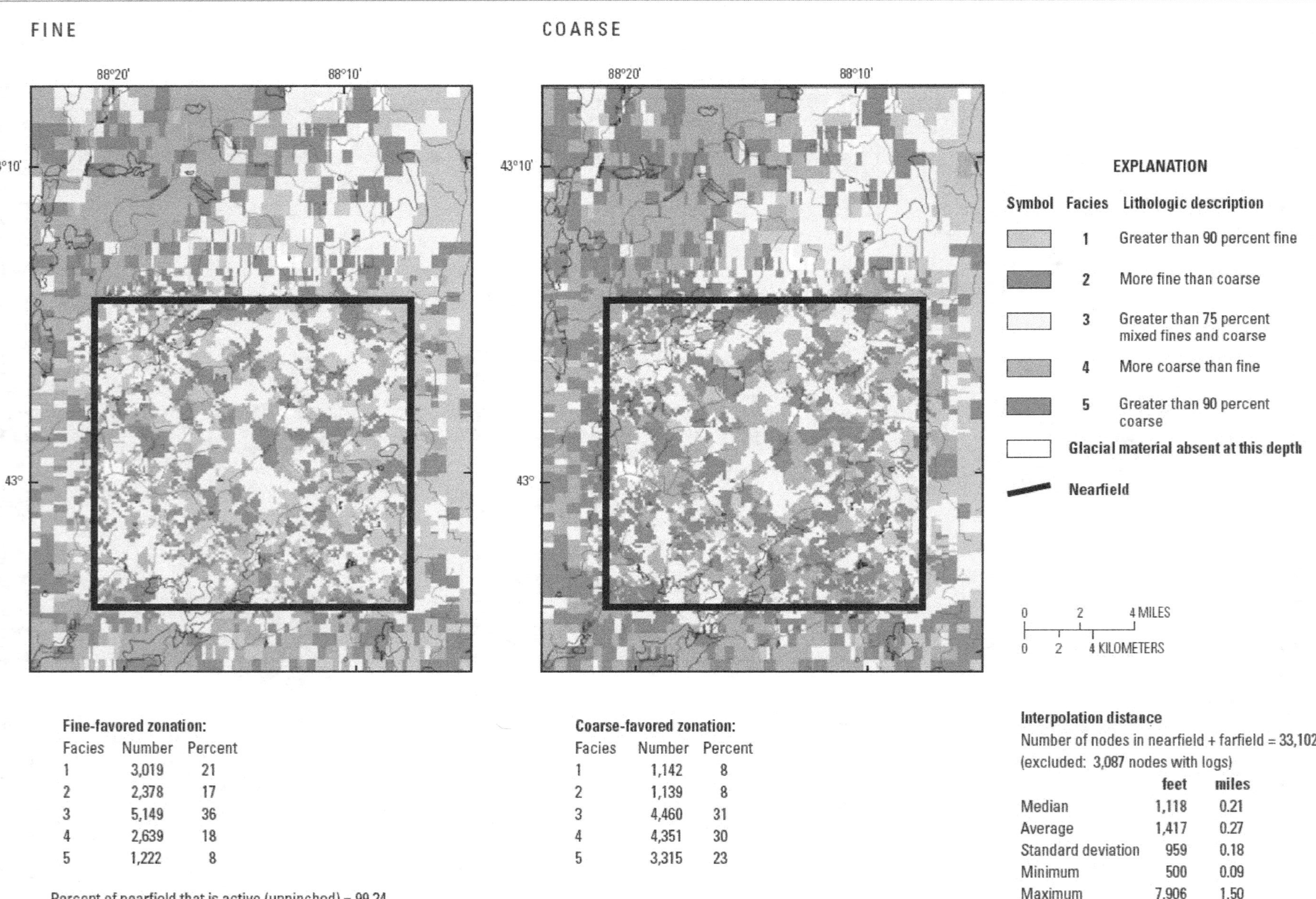

Figure 15A. Fine- and coarse-favored hydraulic conductivity zones by model layer for unconsolidated deposits—layer 1 (0–20 foot depth).

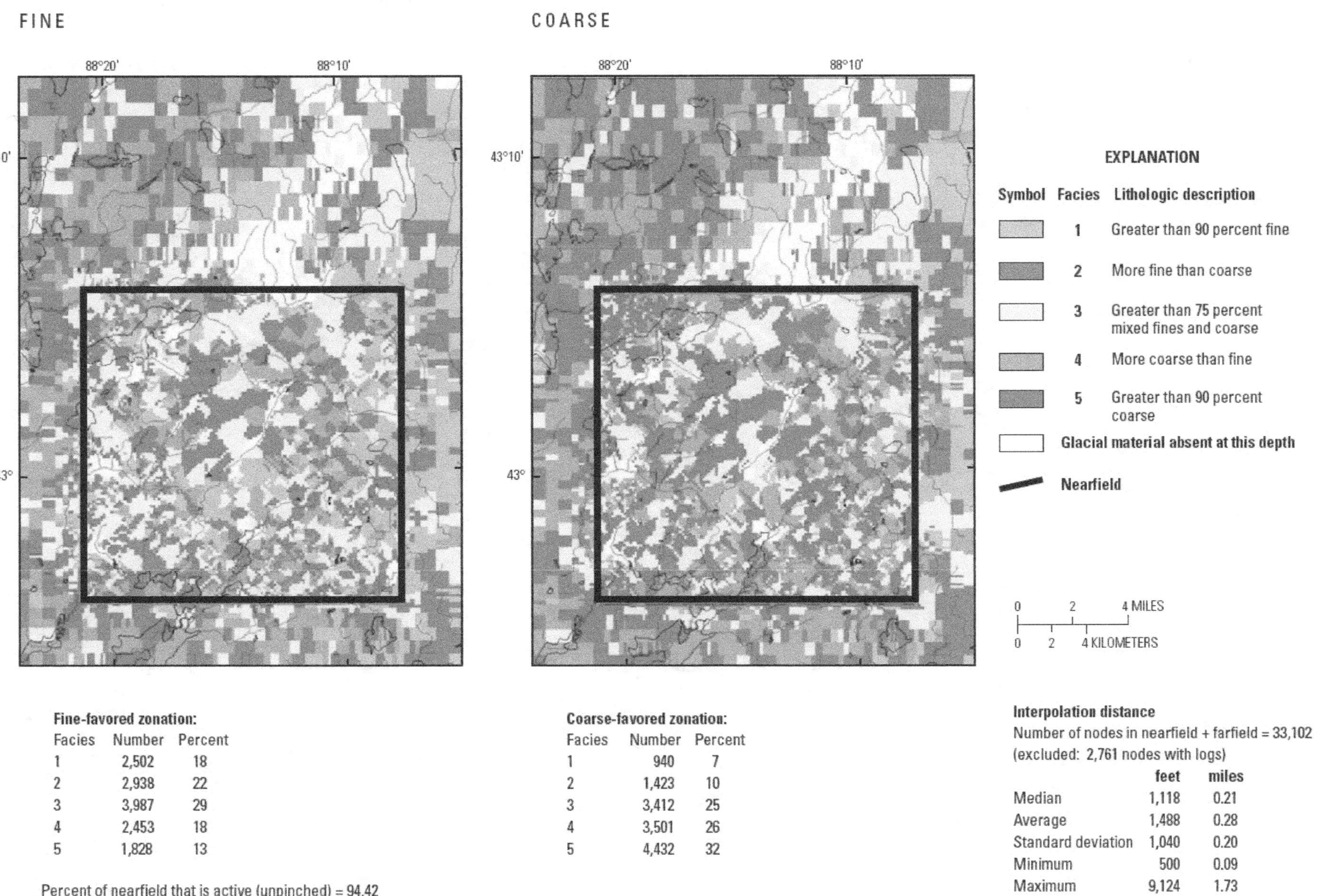

Fine-favored zonation:

Facies	Number	Percent
1	2,502	18
2	2,938	22
3	3,987	29
4	2,453	18
5	1,828	13

Percent of nearfield that is active (unpinched) = 94.42

Coarse-favored zonation:

Facies	Number	Percent
1	940	7
2	1,423	10
3	3,412	25
4	3,501	26
5	4,432	32

Interpolation distance

Number of nodes in nearfield + farfield = 33,102
(excluded: 2,761 nodes with logs)

	feet	miles
Median	1,118	0.21
Average	1,488	0.28
Standard deviation	1,040	0.20
Minimum	500	0.09
Maximum	9,124	1.73

Figure 15*B*. F ne- and coarse-favored hydraulic conductivity zones by model laye for unconsolidated deposits—laye 2 (20–50 foot depth)

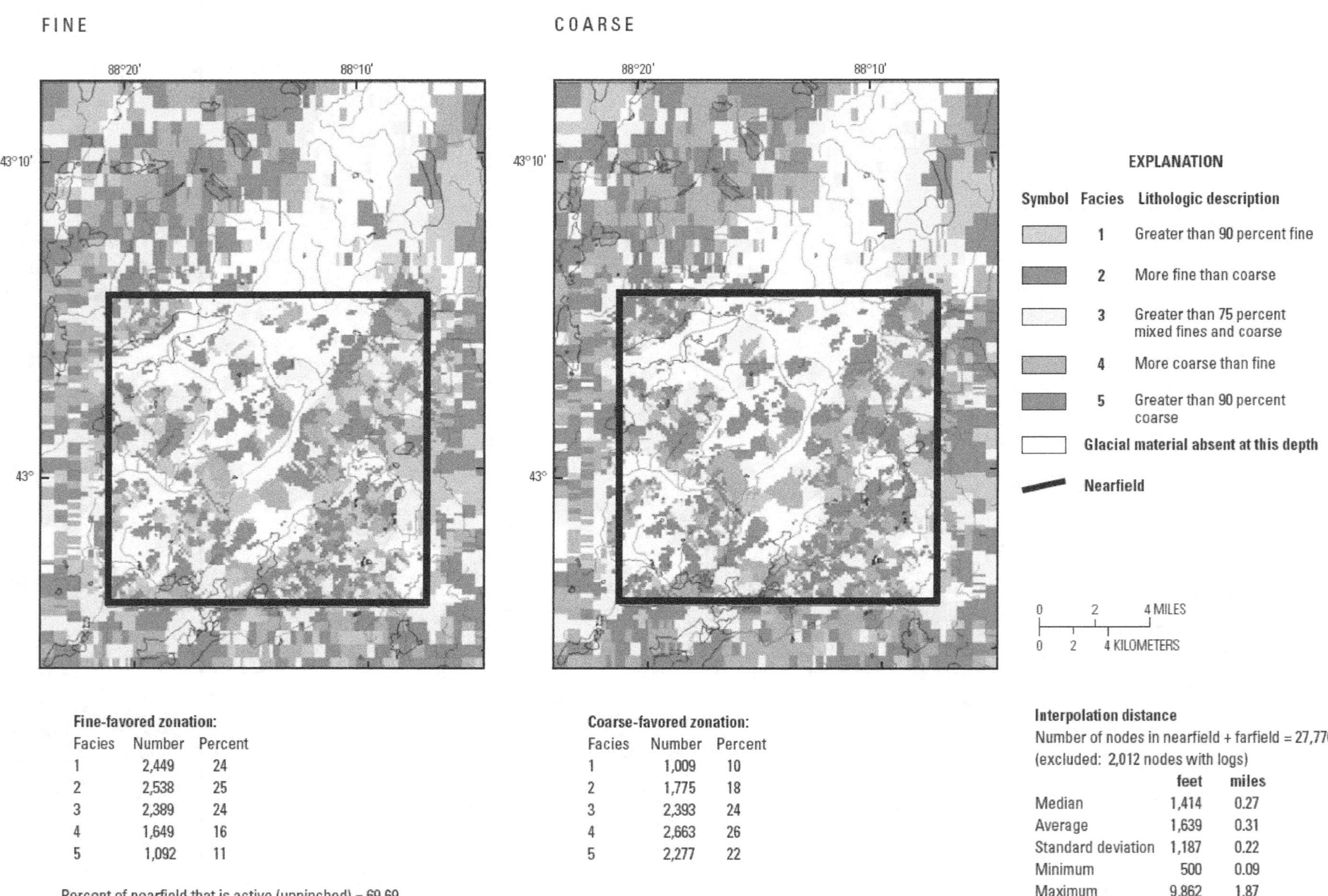

Fine-favored zonation:

Facies	Number	Percent
1	2,449	24
2	2,538	25
3	2,389	24
4	1,649	16
5	1,092	11

Percent of nearfield that is active (unpinched) = 69.69

Coarse-favored zonation:

Facies	Number	Percent
1	1,009	10
2	1,775	18
3	2,393	24
4	2,663	26
5	2,277	22

Interpolation distance

Number of nodes in nearfield + farfield = 27,776 (excluded: 2,012 nodes with logs)

	feet	miles
Median	1,414	0.27
Average	1,639	0.31
Standard deviation	1,187	0.22
Minimum	500	0.09
Maximum	9,862	1.87

Figure 15C. Fine- and coarse-favored hydraulic conductivity zones by model layer for unconsolidated deposits—layer 3 (50–100 foot depth)

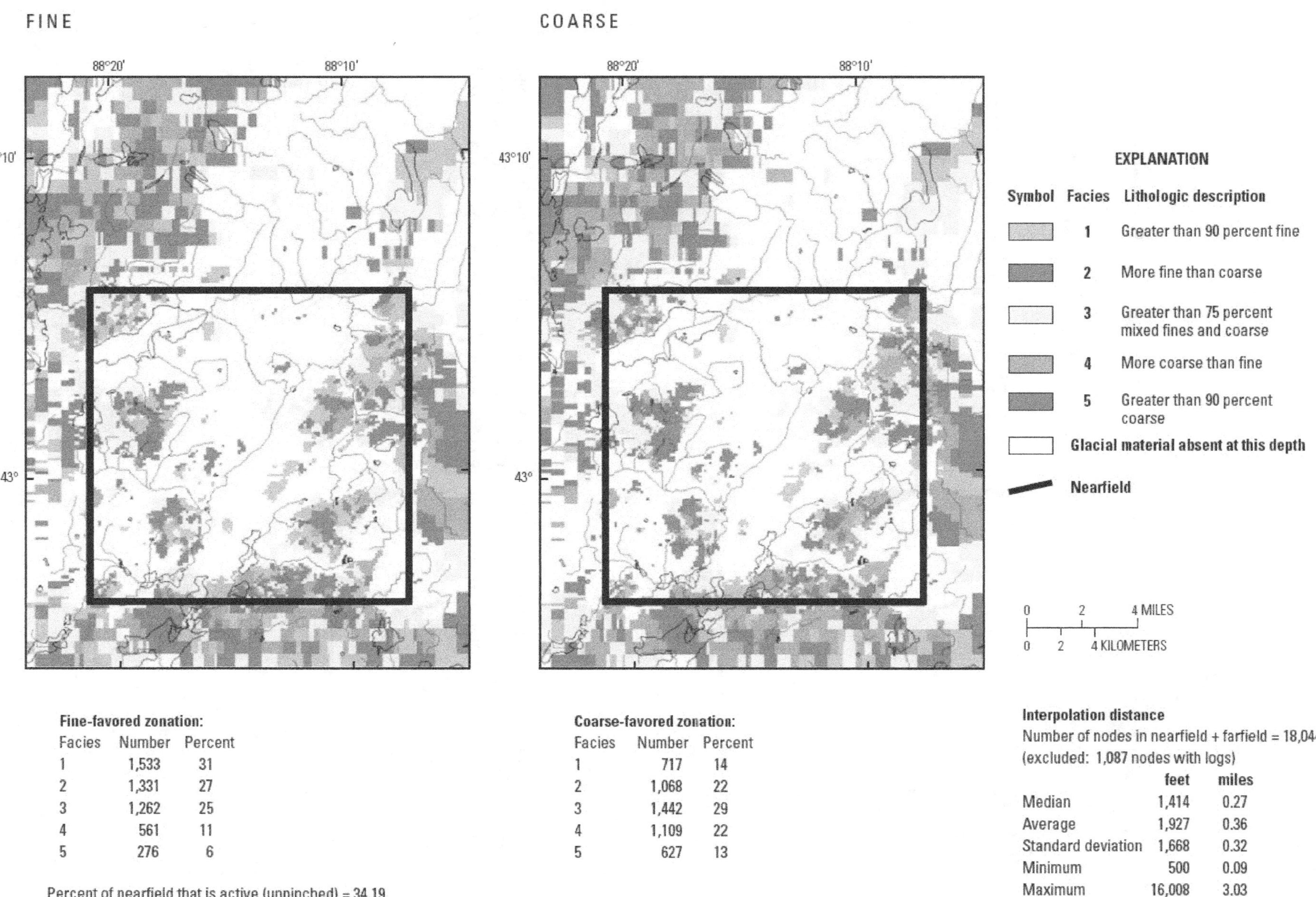

Fine-favored zonation:

Facies	Number	Percent
1	1,533	31
2	1,331	27
3	1,262	25
4	561	11
5	276	6

Percent of nearfield that is active (unpinched) = 34.19

Coarse-favored zonation:

Facies	Number	Percent
1	717	14
2	1,068	22
3	1,442	29
4	1,109	22
5	627	13

Interpolation distance

Number of nodes in nearfield + farfield = 18,044
(excluded: 1,087 nodes with logs)

	feet	miles
Median	1,414	0.27
Average	1,927	0.36
Standard deviation	1,668	0.32
Minimum	500	0.09
Maximum	16,008	3.03

Figure 15*D*. Fine- and coarse-favored hydraulic conductivity zones by model layer for unconsolidated deposits—layer 4 (100–150 foot depth)

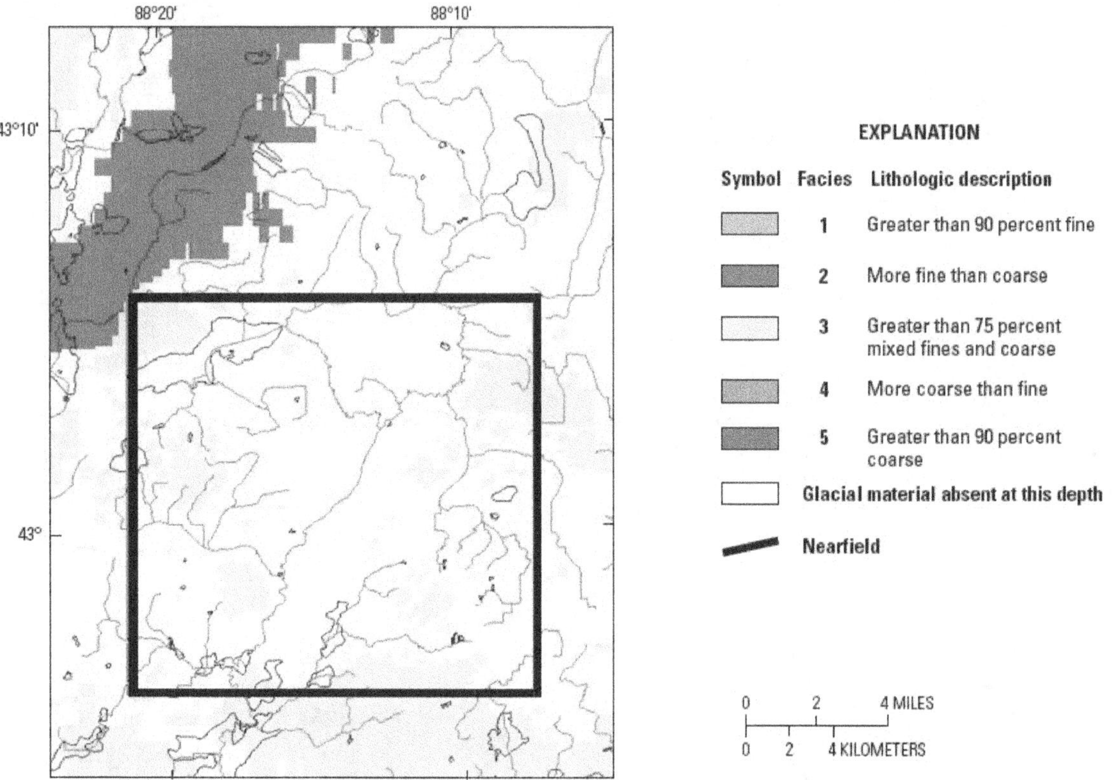

Figure 15*E*. Fine- and coarse-favored hydraulic conductivity zones by model layer for unconsolidated deposits—layer 5 (more than 150 foot depth).

A vertical section can be misleading and tends to underestimate the possibility for preferential flow because it represents only two spatial dimensions (in the case of figure 14, west to east and vertical). Preferential flow along coarse horizons can occur along any direction. The presence of locally continuous, often nonlinear, high-K features in figure 15 suggests (1) that glacial processes of deposition and erosion left some preferential channels intact marked by elevated horizontal hydraulic conductivity (K_h), (2) that these local preferential pathways are most common in the nearfield of the model corresponding to the area of New Berlin Member deposition, and (3) that the pathways are also represented as more common in the coarse-favored than in the fine-favored cases. Pockets of possible preferential vertical flow can be identified by isolating areas where the entire unconsolidated thickness is dominantly coarse (fig. 16). The resulting maps, for a minimum threshold of 60 ft of total unconsolidated thickness, suggest that such areas are more extensive in the coarse-favored (4.6 percent of the nearfield area) than in the fine-favored cases (1.2 percent of the nearfield area) but occupy a small percentage of the nearfield area in both cases.

In summary, the fine-favored and coarse-favored interpretations of the unconsolidated hydrogeology are the basis for two distinct (although related) models of the Upper Fox River Basin—one with hydraulic conductivity zones that favor the connectivity of fine-grained deposits and a second with zones that favor the connectivity of coarse-grained deposits.

3.2 2 Silurian Bedrock

As noted previously, the Silurian dolomite is assumed to be zoned vertically with respect to hydraulic conductivity. The upper 20 ft of the dolomite (layer 6) is taken to represent a weathered horizon and is assigned a higher hydraulic conductivity than the remaining thickness (layer 7). The dolomite as a whole can be divided into several units with differing properties (Rovey, 1990; Carlson 2001). However, for this study, which is focused on the overlying unconsolidated material, the Silurian units are lumped into a single aquifer whose average hydraulic conductivity is assumed to be homogeneous except in areas adjacent to bedrock valleys where the glaciers excavated the dolomite and favored the formation of weathered zones that worked laterally into the rock (Feinstein and others, 2005a). The bedrock valleys correspond to areas of thin or absent dolomite (see fig. 4*B*).

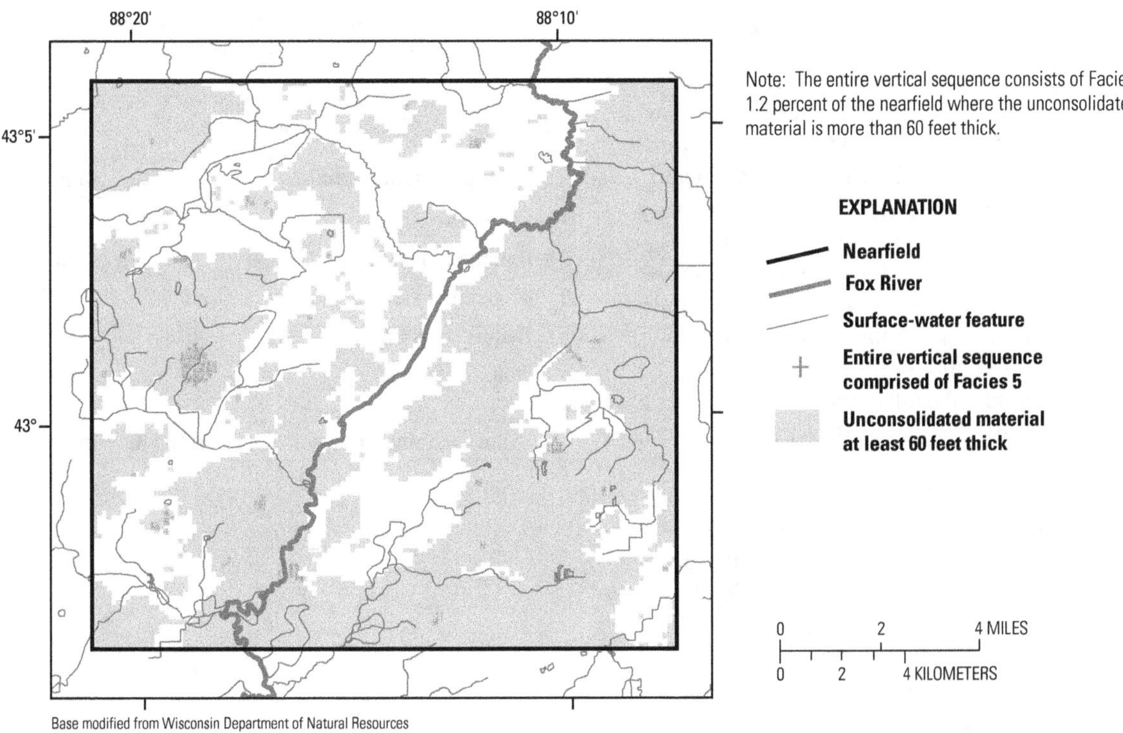

Figure 16*A*. Areas where vertical sequence consists entirely of dominantly coarse facies—fine-favored model. (Well-log data interpreted from Wisconsin Department of Natural Resources, 2009, and Wisconsin Geological and Natural History Survey, 2004, digital-data sources.)

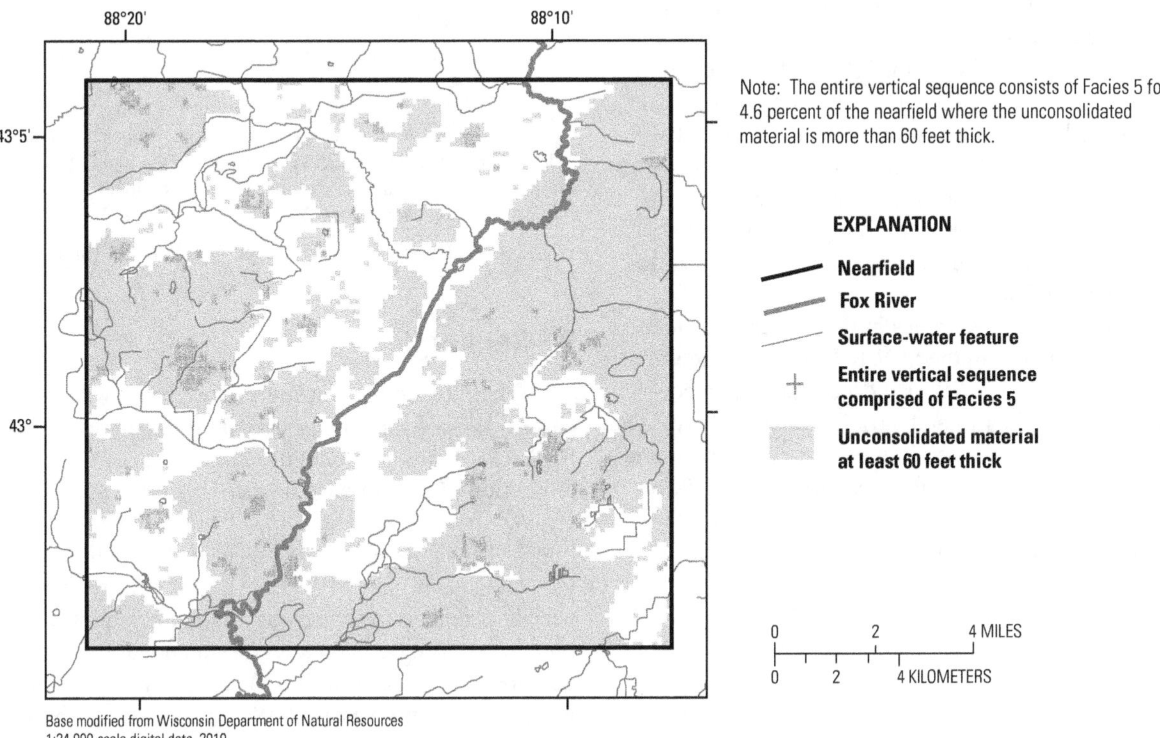

Figure 16*B*. Areas where vertical sequence consists entirely of dominantly coarse facies—coarse-favored model. (Well-log data interpreted from Wisconsin Department of Natural Resources, 2009, and Wisconsin Geological and Natural History Survey, 2004, digital-data sources.)

3.2.3 Initial Hydraulic Conductivity Values

Hydraulic conductivity values in the SEWRPC model were calibrated to conditions across southeastern Wisconsin (Feinstein, and others, 2005a). The K_h values for the regional model vary between 0.2 and 100 ft/d for the unconsolidated material, are set to 4 ft/d in the weathered dolomite, and vary between 1 and 4 ft/d for the remaining dolomite thickness (with the higher K_h values near the Silurian dolomite subcrops adjacent to bedrock valleys). The K_v values for the regional model vary between 0.005 ft/d and 1 ft/d for the unconsolidated material, are set to 0.01 ft/d for the weathered dolomite, and vary between 0.001 and 0.01 ft/d for the remaining dolomite thickness (with the higher K_v values near the Silurian dolomite subcrops adjacent to bedrock valleys).

The calibrated values from the SEWRPC model are the basis for the choice of initial hydraulic conductivity values for the Upper Fox River Basin models. In the case of the unconsolidated material, five K zones are required for the fine-favored and the coarse-favored models. The K_h and K_v values for the finer facies correspond to the low end of the range adopted in the SEWRPC model, and the values for the coarser facies correspond to the high end of the SEWRPC range. The distribution of initial K values by facies is presented in table 4. An important feature of the initial assignments is the implied K_h to K_v anisotropy ratio. It is higher for facies 1 to 4 than it is for facies 5 because facies 1 to 4 are considered to contain at least some fine-grained horizons that inhibit vertical flow, whereas facies 5, dominantly coarse, is represented as largely free of fine-grained horizons.

It bears emphasizing that the **initial K values** are identical for the fine-favored and coarse-favored models. Indeed, the only difference between the initial versions of these two models is the **K zonation** of the facies across layers, as shown by the contrasting plots in figure 15. During the calibration process, the K values are allowed to diverge as do several other input parameters that yields distinct parameterizations for the final versions of the two models.

The Silurian dolomite K values in the Upper Fox model correspond to the zonation in the SEWRPC model. The initial K_h and K_v values assigned to Silurian dolomite layer 6 correspond to zone 8 in table 4 (4 and 0.01 ft/d, respectively). The initial values assigned to most of dolomite layer 7 correspond to zone 1 in table 4 (1 and 0.001 ft/d, respectively); however values increase gradually to the weathered zone 11 values toward the margins of bedrock valleys.

Table 4. Initial hydraulic conductivity values for both fine-favored and coarse-favored models.

[K_h (ft/day), horizontal hydraulic conductivity; K_v (ft/day), vertical hydraulic conductivity; Aniso, anisotropic ratio of K_h to K_v]

Unconsolidated facies	Initial K_h	Initial K_v	Aniso
1 = Dominantly fine	0.5	0.005	100:1
2 = Relatively fine	2	.01	200:1
3 = Mixed fine and coarse	10	.02	500:1
4 = Relatively coarse	40	.2	200:1
5 = Dominantly coarse	80	2.0	40:1

Silurian dolomite zones[1]	Initial K_h	Initial K_v
1	1.00	0.001
2	1.25	.002
3	1.75	.003
4	2.25	.004
5	2.75	.005
6	3.25	.006
7	3.75	.007
8	4.00	.01

Streambed[2]	Initial K_v
Low (characterized as silty/muddy)	1.0
Not characterized	5.0
High (characterized as sandy/gravelly)	25.0

[1] The mapping of the Silurian dolomite zones in layers 6 and 7 of the Upper Fox River Basin model corresponds to the input of the Silurian layers in the regional southeastern Wisconsin regional groundwater-flow model (Feinstein and others, 2005a).

[2] Streambed characterization derived from Poff and Threinen (1963).

3.3 Recharge

The most important source for groundwater in southeastern Wisconsin is natural recharge to the water table. The SEWRPC regional model used recharge rates estimated as a function of stream base flow for watersheds in southeastern Wisconsin (Cherkauer, 2004; Cherkauer and Ansari, 2005). The approach correlates base flow estimated for gaged basins (assumed to be as a proxy for recharge) with variables related to topography, climate, land use, and soil type in order to develop a regression equation applicable to neighboring ungaged basins. The underlying method, detailed in Cherkauer and Ansari, 2005, equates stream base flow to recharge (R), which is normalized to observed annual precipitation (P). Regression analysis shows that R/P is controlled by three dimensionless ratios: (1) infiltrating to overland water flux, (2) vertical to lateral distance water must travel, and (3) natural to developed land cover. The individual watershed properties that indicate or comprise these ratios are commonly available in Geographic Information System (GIS) databases. The empirical relation for predicting R/P was first developed for selected watersheds and then tested outside the study area against other methods for calculating recharge. The method produces values that agree with base-flow separation from streamflow hydrographs (to within 15 to 20 percent), groundwater budget analysis (4 percent), well hydrograph analysis (12 percent), and the distributed-parameter watershed model, PRMS, calibrated to total streamflow (18 percent).

The recharge rates produced by the method are estimates of long-term average rates (rather than rates associated with any given year such as 2005). It yields a mean recharge rate for the region of southeastern Wisconsin equal to 4.5 inches/yr (Feinstein and others, 2005a). For this study, the recharge regression equation developed for southeastern Wisconsin was applied to 27 drainage subbasins within the model domain (Professor Douglas Cherkauer, University of Wisconsin-Milwaukee, written commun., December 6, 2009). The resulting rates vary between 1.6 and 9.5 inches per year (in/yr) (fig. 17). Note, however, that zero recharge is assigned to areas covered by Pewaukee Lake and Vernon Marsh.

The recharge rates are partly correlated to the type of glacial material (mapped in fig. 2). The lowest rates tend to be in the eastern part of the model domain (in areas underlain by the Oak Creek Formation) and the highest rates in the western part (in areas underlain by the Horicon Member). Recharge rates are most variable in the central part of the study area (mostly underlain by the New Berlin Member). The average recharge rates for the model domain and nearfield are 4.43 and 4.16 in/yr, respectively. It is worth noting that although the recharge zonation in the parent SEWRPC model derived from the regression equation was based on larger basins than those used for mapping recharge in the Upper Fox model, the average rate in the part of the SEWRPC model corresponding to the Upper Fox model domain was also 4.43 in/yr, implying that the boundary fluxes transferred from the SEWRPC model to the Upper Fox model are consistent with the recharge rates applied inside the Upper Fox model.

3.4 Surface-Water Network

The surface-water network for the model domain (fig. 6*A*) consists of streams, lakes, and wetlands. These features are represented by four MODFLOW packages (fig. 18), two of which are applied outside the Upper Fox River Basin (RIV and DRN) and three inside the Basin (SFR, LAK and DRN).

3.4.1 Outside Upper Fox River Basin

Streams outside the Upper Fox River Basin are represented by the RIV package and water bodies (lakes and wetlands) are represented by the DRN package. From an input standpoint, these packages are populated by relatively simple inputs—estimates of stage and estimates of streambed conductance—which in turn are a function of the streambed K_v, thickness, width, and length inside a model cell. Base flow to a stream occurs from the groundwater to the stream in accordance with Darcy's Law whenever the simulated water-table elevation is higher than the prescribed stream stage in a cell. The RIV package also allows loss of water from the stream to the groundwater when the stream stage is higher than the ambient water table. Maximum inflow from the stream to the groundwater is reached when the water-table elevation is at or below the bottom of the streambed elevation. The DRN package only allows outflow from groundwater to the simulated water body. By assigning lake and wetlands to the DRN package, the model does not allow water bodies that are simulated as perched above the water table from influencing the model solution.

3.4.2 Inside Upper Fox River Basin

Streams inside the Upper Fox River Basin are represented by the SFR package and water bodies are represented by the LAK or DRN package. The SFR algorithm in MODFLOW permits more sophisticated simulation of groundwater/surface-water interactions than does the RIV package (Niswonger and Prudic, 2006). The streamflow in the channel is propagated downstream through the network from lower order to higher order tributaries. For example, the amount of simulated flow is calculated in such a way that if a well is pumping at a rate high enough to dewater a part of the channel, the water available to the well from the stream will be limited. In the Upper Fox River Basin models, the streamflow in the channel is coincident with the contribution from accumulated base flow (including inflows from the Sussex, Brookfield, and Waukesha WWTPs) , but the overland runoff component is not considered. Consequently, the model output for streamflow corresponds conceptually to low flow conditions typical of late summer or early fall when runoff is minimal. Because the entire Upper Fox River Basin is incorporated in the model domain, including the headwaters of all tributaries to the Fox, there is no need to add water to channels to account for accumulated base flow originating outside the model.

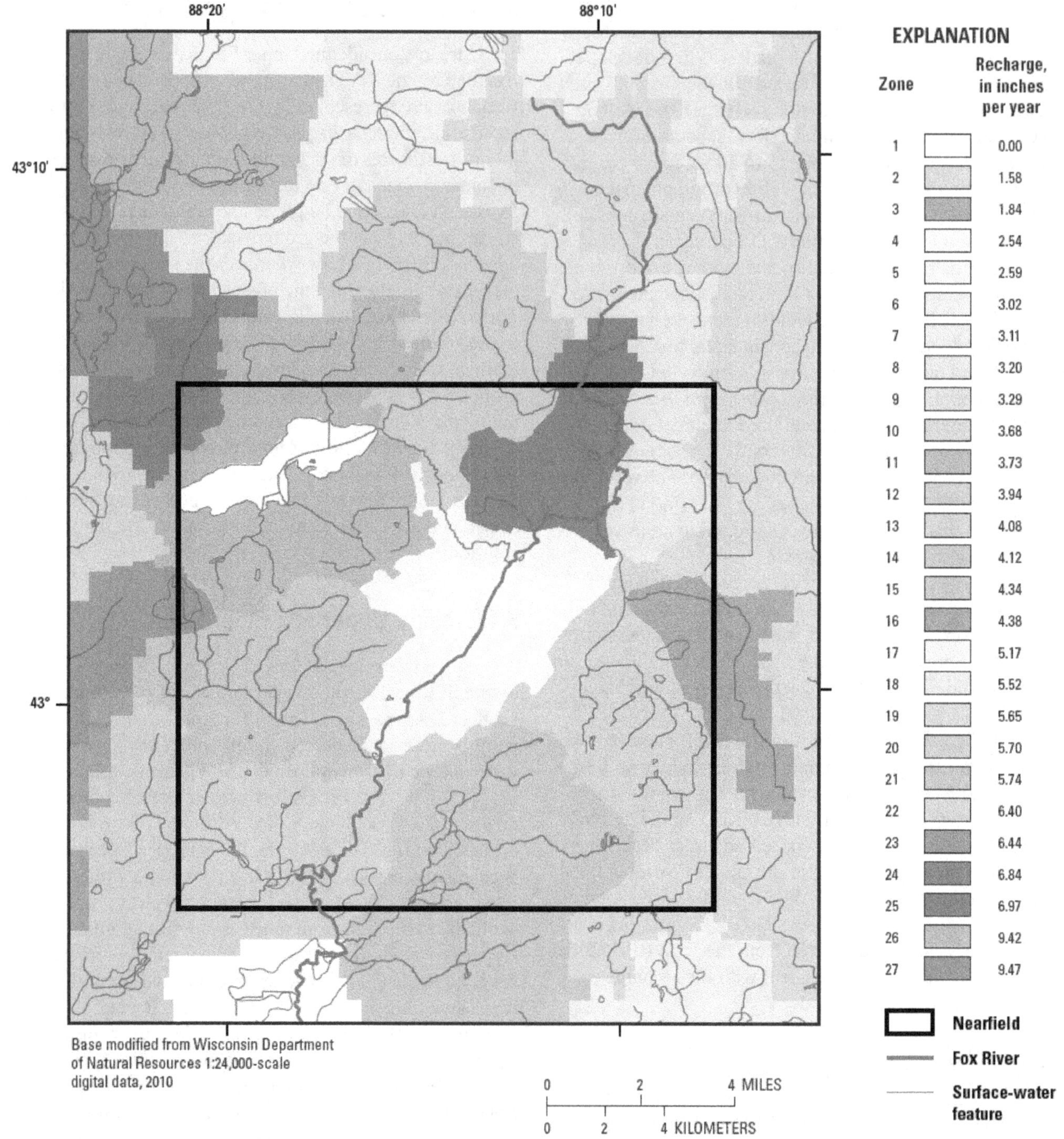

88°20'
88°10'
43°10'
43°

EXPLANATION

Zone	Recharge, in inches per year
1	0.00
2	1.58
3	1.84
4	2.54
5	2.59
6	3.02
7	3.11
8	3.20
9	3.29
10	3.68
11	3.73
12	3.94
13	4.08
14	4.12
15	4.34
16	4.38
17	5.17
18	5.52
19	5.65
20	5.70
21	5.74
22	6.40
23	6.44
24	6.84
25	6.97
26	9.42
27	9.47

Nearfield
Fox River
Surface-water feature

Base modified from Wisconsin Department of Natural Resources 1:24,000-scale digital data, 2010

0 2 4 MILES
0 2 4 KILOMETERS

Figure 17. Model recharge distribution. (Recharge values based on Cherkauer and Ansari, 2005, p. 102–112.)

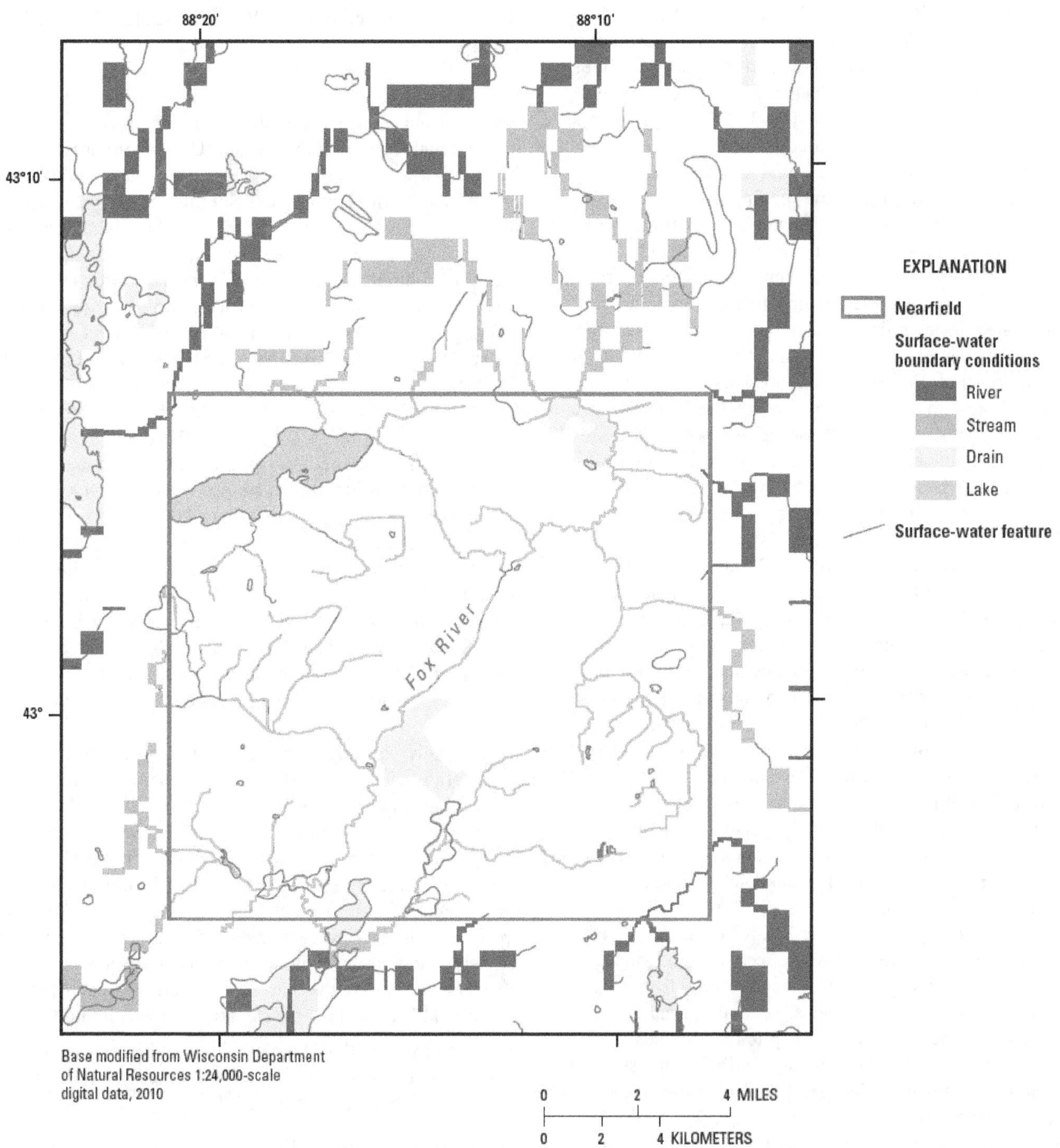

Figure 18. Surface-water cells assigned to MODFLOW-2005 packages.

Another capability of the SFR package is to calculate stream stage (rather than accept it as an input) by virtue of the net accumulated upstream base flow and the wetted channel geometry, which, through the channel width, also is partly a function of the accumulated base flow. In this study, the capability to calculate stage is only used for the headwater reaches of the stream network inside the Fox River Basin; that is, for the "first order" reaches above the first confluence with a second tributary. If the model simulates a water-table elevation below the top of the streambed for a headwater reach, then the reach is reported dry and is, by implication, ephemeral. If the model accumulates base flow in a headwater reach, then it is simulated as perennial. About one-half the total length of streams in the Upper Fox River Basin models corresponds to headwater reaches; the reaches that are second order or above (and are unlikely to be ephemeral) are assigned a fixed stage. Like the RIV package, the SFR package allows water to flow from the stream channel to the groundwater when the simulated water table is below the stage (fixed or calculated).

The Upper Fox River Basin contains one large water body—Pewaukee Lake. This feature is simulated with the LAK package (Merritt and Konikow, 2000) that simulates lake level in terms of the lake geometry and the balance of inflow and outflow of water. The water budget for the lake takes account of precipitation on the lake surface, any runoff from the lake basin area, direct groundwater base flow, and tributary streamflow as inflows. The budget also takes account of evaporation from the lake surface, losses to groundwater, streamflow into an outlet, and any direct pumping of lake water as outflows. In the case of Pewaukee Lake, direct runoff is neglected (as it is for the stream network) and no pumping is reported to occur, but the other budgets terms, including contributions from inflowing streams (such as Zion Creek) and discharge to an outlet stream (Pewaukee River and then to the Fox River) are simulated.

The topography in the Upper Fox River Basin includes valleys and lowlands (fig. 5*B*) where wetlands in connection with the groundwater system are present. These features are often in close proximity to streams. Several riparian wetlands associated with the Fox River are explicitly simulated in the models by means of the DRN package with the largest occurring near the confluence of Sussex Creek with the Fox River, upstream from the confluence of Pebble Creek with the Fox, and just north of Vernon Marsh where Pebble Brook joins the Fox. Most of the groundwater discharge to these riparian features is expected to flow a short distance overland into the river channel and, therefore, for the purposes of model accounting, the DRN flow to these features is added to the accumulated base flow in the Fox River.

3.4.3 Input to MODFLOW-2005 Surface-Water Packages

The mapping of the surface-water network was constructed from spatial information available in GIS format in the National Hydrography Dataset (U.S. Geological Survey, 2010b) that was clipped to the model study area and overlain on the model grid. Each model cell intersected by a surface-water feature requires a number of inputs to characterize the connection of the feature to the groundwater system. An input common to the RIV, DRN, and SFR packages for second order and higher streams is the surface-water stage (whereas the stage is calculated for headwater SFR streams and for the application of the LAK package to Pewaukee Lake). The fixed stages are linked to the land-surface elevations derived from Waukesha County and USGS DTM files discussed in the "Model Construction" section under "Vertical Layering". To interpolate from the equally-spaced estimates of land-surface elevation to the midpoints of stream stretches inside model cells, a triangular irregular network (TIN) consisting of the three land-surface elevations points nearest the stage location was constructed and the stage elevation determined using weights calculated from the natural neighbor algorithm referenced in the "Vertical Layering" subsection Each resulting stage elevation was inspected to ensure that stream elevations decreased in the downstream flow direction and, if not, a linear correction was applied across the reach to enforce downstream routing.

A likely bias arising from this procedure is estimated stages are too high given that the TIN surface is unlikely to reflect channel depressions and is more likely to reflect the elevation of the first terrace bordering a stream. To account for the bias, a comparison was made between the TIN-derived stages and stages recovered from USGS 7.5-minute topographic maps (scale: 1 inch equals 2,000 ft) in the Waukesha area. Comparison of elevations were calculated as interpolated TIN elevation minus contour elevation at about 350 locations (fig. 19) where the map contours cross streams (varying between 717 ft and 1,069 ft elevation). It was found that:

- on average the TIN elevations are higher than the contour elevations; the mean difference is +3.2 ft and the median difference is +2.7 ft;
- 49 percent of the differences are between +3 and −3 ft and 83 percent are between +6 and −6 ft, but only 2.8 percent of the comparisons are more negative than −3 ft;
- the biggest differences are almost always associated with headwater reaches (fig. 19).

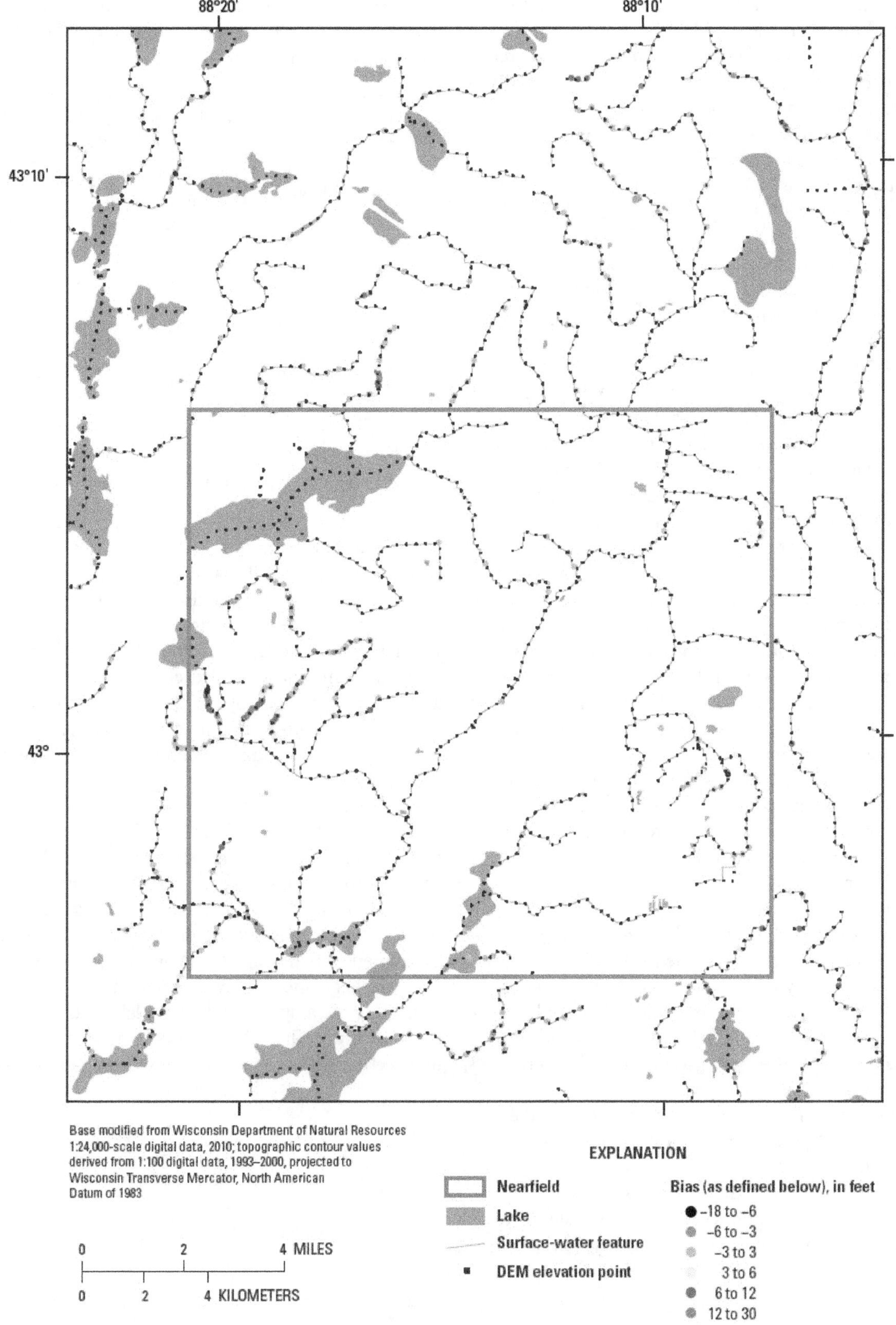

Figure 19. Comparison of topographic contours to digital elevation model (DEM) (LIDAR) elevations at stream locations. Biases calculated by subtracting the elevation where a topographuic contour crosses a stream from the elevation obtained from the interpolated DEM at the same location.

To correct the bias, two actions were taken. All TIN-generated stream elevations in the farfield and the nearfield were lowered by 2.7 ft (corresponding to the median bias); in areas of large difference, the stream elevations were also manually corrected to equal contour elevations. After the corrections, more than 70 percent of the estimated stages are within 3 ft of the stages implied by the contour maps.

The stages of lakes are directly imported from the original land-surface databases. Little fitting was needed to make the elevations of lakes integrated in the surface-water network consistent with the stage elevations of inlet and outlet streams. The stages of riparian wetlands included as groundwater discharge areas were set 1 ft below the land surface assigned to the occupied model cells.

A second input common to surface-water packages is the conductance term mediating the flux exchange between groundwater and the stream, lake, or wetland. The conductance term is equal to the product of streambed K_v, width, and length inside the model cell divided by assumed streambed thickness. For the streams inside the Upper Fox River Basin represented by the SFR package, a compilation of the properties of surface-water features for Waukesha County (Poff and Threinen, 1963) served to separate the reaches into three zones (fig. 20) based on bed descriptions as silty/muddy (low K_v zone), sandy/gravelly (high K_v zone), or the absence of a description (middle K_v zone). The bed of the Fox River falls into the unknown, that is middle, category. To quantify the K_v values , notice was taken of a study conducted in Dane County (located about 50 mi west of the study area), where measured riverbed K_v values (assuming bed thickness of 1 ft) ranged between 1.6 ft/d and 37 ft/d, averaging 8 ft/d (Krohelski and others, 2000). For this study, 1 ft/d was assigned as the initial K_v for the low zone, 5 ft/d was assigned for the middle zone, and 25 ft/d for the high zone (table 4).

A simple equation (table 5) was used to correlate the width component of the Upper Fox stream conductances with the upstream length of the longest trunk of the stream upgradient from the model cell in question. The intent of the equation is to assign widths closely corresponding to the trial widths listed in table 5 for selected upstream lengths. The application of this equation assigns widths less than 8 ft for reaches less than 1 mi in length, typically those associated with headwaters. For larger features, the width can exceed 100 ft as is observed for the Upper Fox River toward its entry into the Vernon Marsh at the downstream end of the model nearfield. The stream length corresponds to the superposition of the GIS National Hydrography Dataset arc coverages on the model grid. Finally, the thickness of the streambed was assumed to be equal to 1 ft for all Upper Fox River Basin stream reaches.

Table 5. Correlation of channel width with upstream length by means of fitting curve for streamflow-routing (SFR) cells in Upper Fox River Basin.

[The main trunk of the Fox River near the southern edge of the nearfield has an upstream length of about 24 miles, which implies a fitted width of 99 feet, roughly equal to the observed channel width at that location, equal to 89 feet]

Upstream length (miles)[1]	Target channel width (feet) assumed for upstream length	Fitted channel width from equation (feet)[2]
0.1	1	1.3
.3	3	3.1
1	8	7.9
5	30	28.2
10	50	48.7
25	100	100.2

[1]Upstream length corresponds to length of main trunk of each stream.

[2]Fitting equation: **Width_ft=0.0092*(Upstream_Length_ft)**$^{0.7884}$.

Inputs particular to the headwater (first order) reaches represented by the SFR package are the slope and Manning coefficients of the streambeds that are used in the calculation of the stream stage by means of the Manning equation (Niswonger and Prudic, 2006). In the absence of data to quantify these inputs for individual reaches in the domain, the models apply a uniform slope of 0.002 ft/ft corresponding to a moderate slope, and a uniform dimensionless Manning coefficient of 0.037 corresponding to a bed with some gravel and cobbles (Barnes, 1967).

For streams outside the Upper Fox River Basin represented by the RIV package, the bed K_v is assumed to be equal to 1 ft/d for all reaches; the stream width for these features is set to 30 ft, the length corresponds to the intersection of GIS arcs with the model grid; and the bed thickness is uniformly set to 1 ft. For lakes outside the Upper Fox River Basin as well as riparian wetlands inside the basin, both represented by the DRN package, the bed Kv is assumed to be equal to 0.01 ft/d (reflecting the assumption that the lake and wetland beds are predominantly silty), the area corresponds to the GIS coverage intersection with model cells, and the bed thickness is set to 2 ft.

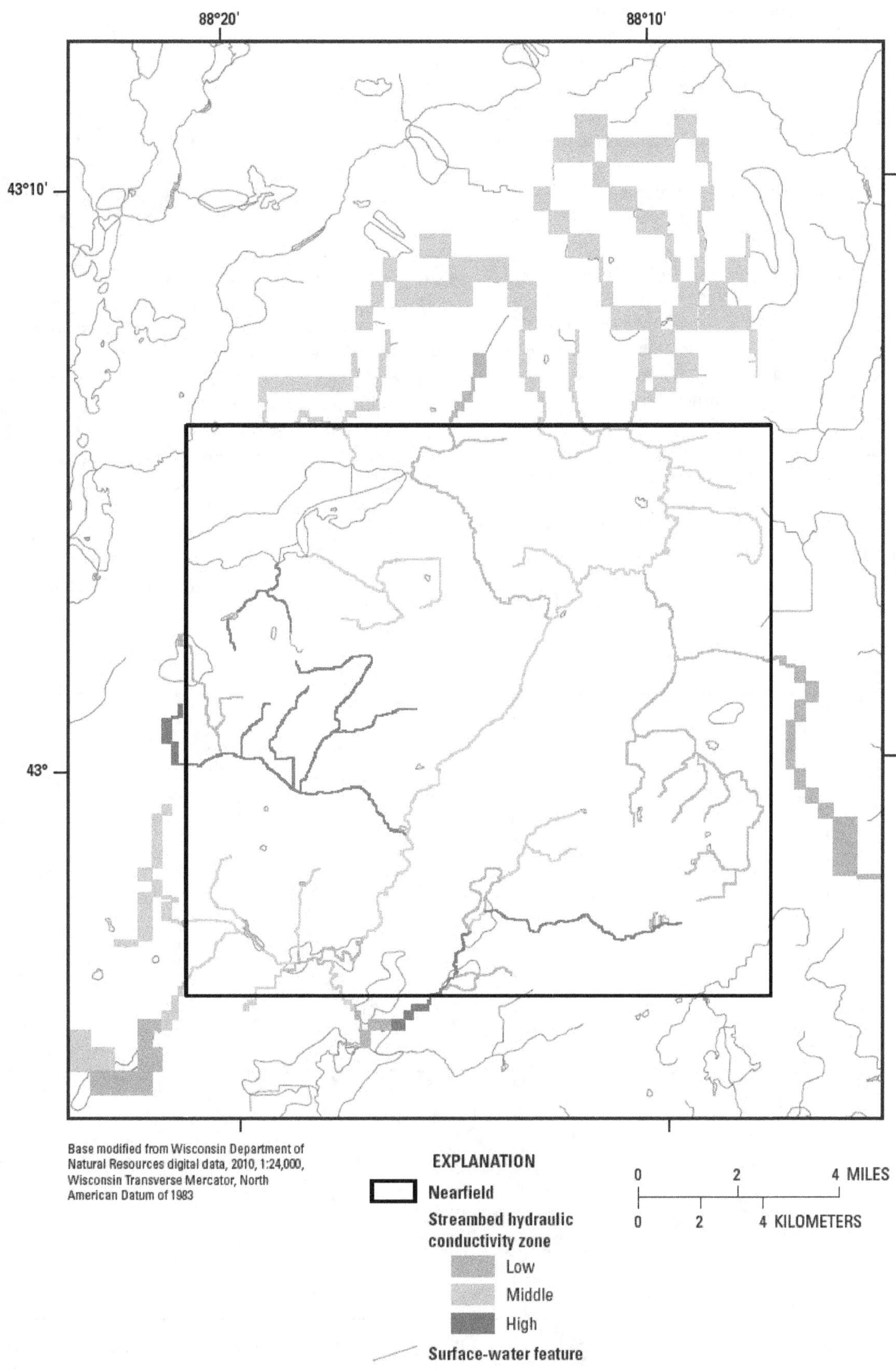

Figure 20. Streambed hydraulic conductivity zones for routed SFR cells.

The LAK package representation of Pewaukee Lake requires special input. Precipitation to and evaporation from the lake surface are assumed to be 32 and 29 in/yr, respectively, on the basis of average long-term observations in northwestern Waukesha County (Linsley and others, 1982, p. 78, 154). Lakebed conductance and thickness are assumed to be equal to 0.01 and 1 ft/d, respectively. Lake stage is a function of inflows, outflows, and the volume of the lake, which in turn depends on the lake stage, lake area, and lake bottom. The geometry and bathymetry of Pewaukee Lake have been mapped (Wisconsin Conservation Department, 1966). The survey indicates that the average area of the lake is on the order of 2,425 acres and the volume is on the order of 35,507 acre-ft. The 7,059 LAK cells (each 125 ft on a side) approximate the area, and a flat bottom assigned to the lake cells set at 738 ft elevation approximates the desired volume for the expected stage between 852.5 and 853 ft.

A key input for calculating lake stage is the relation between lake stage and outflow at the outlet to Pewaukee River. Data made available by the engineer who controls the weir at the spillway outlet (David White, Director of Public Works/Village Engineer, Village of Pewaukee, written commun., September 2009) allows a rating curve to be constructed between lake stage and outflow using the relation from Streeter (1966):

$$Q = 3.09 * w * (h^{1.5}) \tag{2}$$

where

- Q is approximated outflow, in cubic feet per second;
- w is width, in feet, of weir at outlet equals 12; and
- h is calculated lake stage above spillway elevation, in feet.

The spillway elevation varies over time with adjustment of the weir but averaged about 852.35 ft between January 2007 and August 2009. The median recorded difference between the spillway elevation and lake stage (h) was 0.51 ft, with the lake level below the spillway about 20 percent of the time and as much as 3.61 ft above the spillway elevation at maximum lake stage. Applying the weir equation to the period of record and adjusting it to the average spillway elevation yields the relation shown graphically in figure 21. For example, a lake stage of 852.75 ft (corresponding to h equals 0.4 ft) corresponds to an outflow 9.4 cubic feet per second (ft^3/s). Using this relation, MODFLOW-2005 iterates to a lake stage solution reflecting long-term low-flow conditions that balances all inflows and outflows including the discharge across the weir to the Pewaukee River.

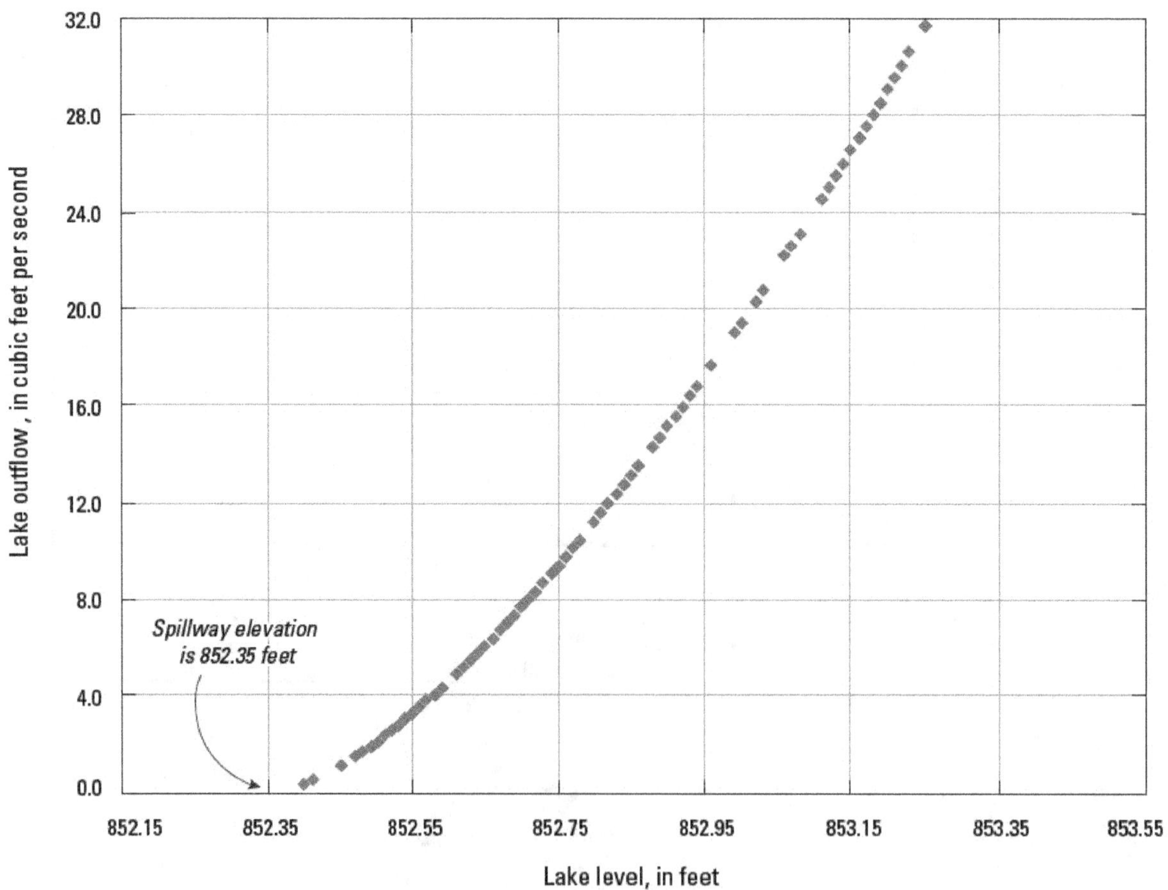

Figure 21. Pewaukee Lake outflow as a function of lake level.

The inputs to the LAK, SFR, RIV, and DRN packages are summarized in table 6. One additional input to the SFR package is the effluent added to the Upper Fox River Basin surface-water network from the three WWTPs. The effluent originates as water pumped from the shallow and deep aquifer systems and is used for public supply and industrial/commercial purposes before treatment and discharge. In low-flow periods, the combined effluent flow represents about one-third of the streamflow in the Upper Fox River above the Vernon Marsh (see "Model Results" section). Effluent flows recorded by the Sussex, Brookfield, and Waukesha water utility agencies at the WWTPs (Professor Timothy Grundl, University of Wisconsin-Milwaukee, written commun., February 15, 2010) show daily and seasonal rates vary around the average as a function of water demand, and there has been a long-term upward trend since the 1960s. Model input flows between September 2008 and July 2009 are averaged to yield a single value at each WWTP—Sussex, 2.44 ft^3/s; Brookfield, 10.85 ft^3/s; and Waukesha, 13.83 ft^3/s. These average rates are added to the model-calculated streamflow at the WWTP locations (fig. 6*B*).

Table 6. Input for MODFLOW surface-water packages.

[NHD, U.S. Geological Survey National Hydrography Dataset (2010a); land-surface dataset from Waukesha County (2005); LAK, lake; ft/d, foot per day; ft, foot; in/yr, inch per year; SFR, streamflow routing; ft/ft; foot per foot; RIV, river; DRN, drain]

	Value or data source
LAK7 package for Pewaukee Lake	
Lakebed hydraulic conductivity	0.01 ft/d
Lakebed thickness	1 ft
Precipitation on lake	32 in/yr
Evaporation from lake	29 in/yr
Runoff to lake	0.0
Pumping from lake	0.0
Lake sill elevation corresponding to outlet into Pewaukee River	852.35 ft
Lake stage/lake discharge relation for outlet	see figure 21
SFR2 package for streams inside Upper Fox River Basin	
Streambed hydraulic conductivity	Zones: low, middle, high (figure 20)
Streambed thickness	1 ft
Channel width in cell	related to upstream length (table 5)
Channel length in cell	NHD coverage intersected with grid
Streambed slope for headwater reaches	0.002 ft/ft
Streambed Manning Coefficient for headwater reaches	0.037
Stream stage in cell for reaches downgradient from headwater reaches	interpolated from land-surface data
RIV package for streams outside Upper Fox River Basin	
Riverbed hydraulic conductivity	1 ft/d
Riverbed thickness	1 ft
Channel width in cell	30 ft
Channel length in cell	NHD coverage intersected with grid
River stage in cell	interpolated land surface
DRN package for water bodies (wetlands and lakes other than Peawaukee Lake)	
Bed hydraulic conductivity	0.01 ft/d
Bed thickness	2 ft
Area in cell	NHD coverage intersected with grid
Stage	land surface for lakes; land surface minus 1 ft for wetlands
DRN package for Sussex and Waukesha Silurian dolomite quarries	
Quarrybed hydraulic conductivity	10 ft/d
Quarrybed thickness	1 ft
Quarry stage	10 ft below top of dolomite in layer 6

3.5 Water Withdrawals

Development of the Upper Fox River Basin has been accompanied by stresses on the groundwater-flow system. The models consider two activities that withdraw groundwater—pumping from high-capacity wells and discharge to dolomite quarries. The addition of water by artificial recharge is not reported for the area and inflow from irrigation, a rare practice in Waukesha County, is neglected.

3.5.1 Pumping

The economic growth and suburban expansion in southeastern Wisconsin have been because of, in part, the abundant water supplies available for public, domestic, and industrial uses. Lake Michigan is the source for about 70 percent of all water used in the region, mainly in the lakeshore counties (Ozaukee, Milwaukee, Racine, and Kenosha), which lie mostly within the Great Lakes drainage basin. Farther inland, Washington, Waukesha, and Walworth Counties are principally in the Mississippi River Basin, and, because of international limitations on diversion of water out of the Great Lakes Basin, rely on groundwater for over 99 percent of their needs (Southeastern Wisconsin Regional Planning Commission and Wisconsin Geological and Natural History Survey, 2002). In 2005, about 95 Mgal/d of groundwater were withdrawn for public, domestic, industrial, commercial, and agricultural uses (Southeastern Wisconsin Regional Planning Commission, 2010). The withdrawal in Waukesha County is estimated to be 34 Mgal/d. About 25 percent of the total is pumped by private domestic wells penetrating shallow aquifers. The remaining withdrawal is extracted from high-capacity wells (defined as withdrawing on average more than 0.1 Mgal/d) penetrating the shallow and deep aquifer systems, chiefly for public supply and industrial purposes.

Recent studies of water use (Buchwald and others, 2010; Feinstein and others, 2010; Southeastern Wisconsin Regional Planning Commission, 2010) indicate that shallow aquifer pumping from high-capacity wells in the model domain for 2005 was on the order of 6.7 Mgal/d. The fine-favored and coarse-favored models include wells open to the unconsolidated material, which pump 1.62 Mgal/d, and wells open to the Silurian dolomite, which pump 5.07 Mgal/d. The greatest concentration of pumping is from dolomite wells in the eastern one-third of the study area (fig. 22). The pumped interval of most wells penetrate multiple model layers, and withdrawal is distributed among the layers according to the relative magnitude of the transmissivity in each layer (equal to the initial hydraulic conductivity values multiplied by the total cell thickness).

Domestic pumping is not included as a model stress (see discussion in "Model Limitations" section). The models also exclude pumping wells installed after 2005—notably three public supply wells located at the south end of the city of Waukesha, which withdraw about 1.6 Mgal/d from the shallow aquifer system (Jeff Detro, Waukesha Water Utility, written commun., May 2010). However, two of these wells are included in the demonstration application of the fine-favored and coarse-favored models used to evaluate the possibility of augmenting water supply by means of riparian pumping (see "Model Application to Hypothetical Well Field" section).

3.5.2 Quarries

The Silurian dolomite is quarried as a construction stone at several sites in Waukesha County. Complexes inside the Upper Fox River Basin include quarries adjacent to Sussex Creek and the Fox River (fig. 5*A*). The excavations remove the unconsolidated overburden and penetrate the top of the bedrock. These features are represented in the model by the DRN package (table 6). At the quarry locations model layers 1 through 5 are pinched (owing to the absence of unconsolidated material), and the drain cells are inserted in model layer 6 at an elevation 10 ft below the top of the bedrock in adjacent unexcavated areas. The conductance term corresponds to the product of the area of the quarry cells and bed K_v is assumed to be equal to 10 ft/d divided by an assumed bed thickness of 1 ft. The resulting conductance value is set deliberately high because little resistance to groundwater discharge is expected through the quarry walls. The fine-favored and coarse-favored models simulate the rate of discharge as part of model output (see "Model Results" section).

3.6 Time Discretization

The base model is a steady-state solution corresponding to recent flux conditions and, therefore, model construction does not involve changes to input over time. However, as part of the calibration process, a pumping test is simulated based on 2005 initial conditions. The transient simulation involves a 2-year runup to the pumping test (during which flow conditions responded to pumping from two wells installed in 2006), followed by the 20-hour test period (during which a third new well was introduced and simulated drawdown around the new well was compared to observed drawdown). The details of the transient simulations of the pumping test for the fine-favored and coarse-favored models are given in "Model Calibration" section under "Aquifer Test Drawdown."

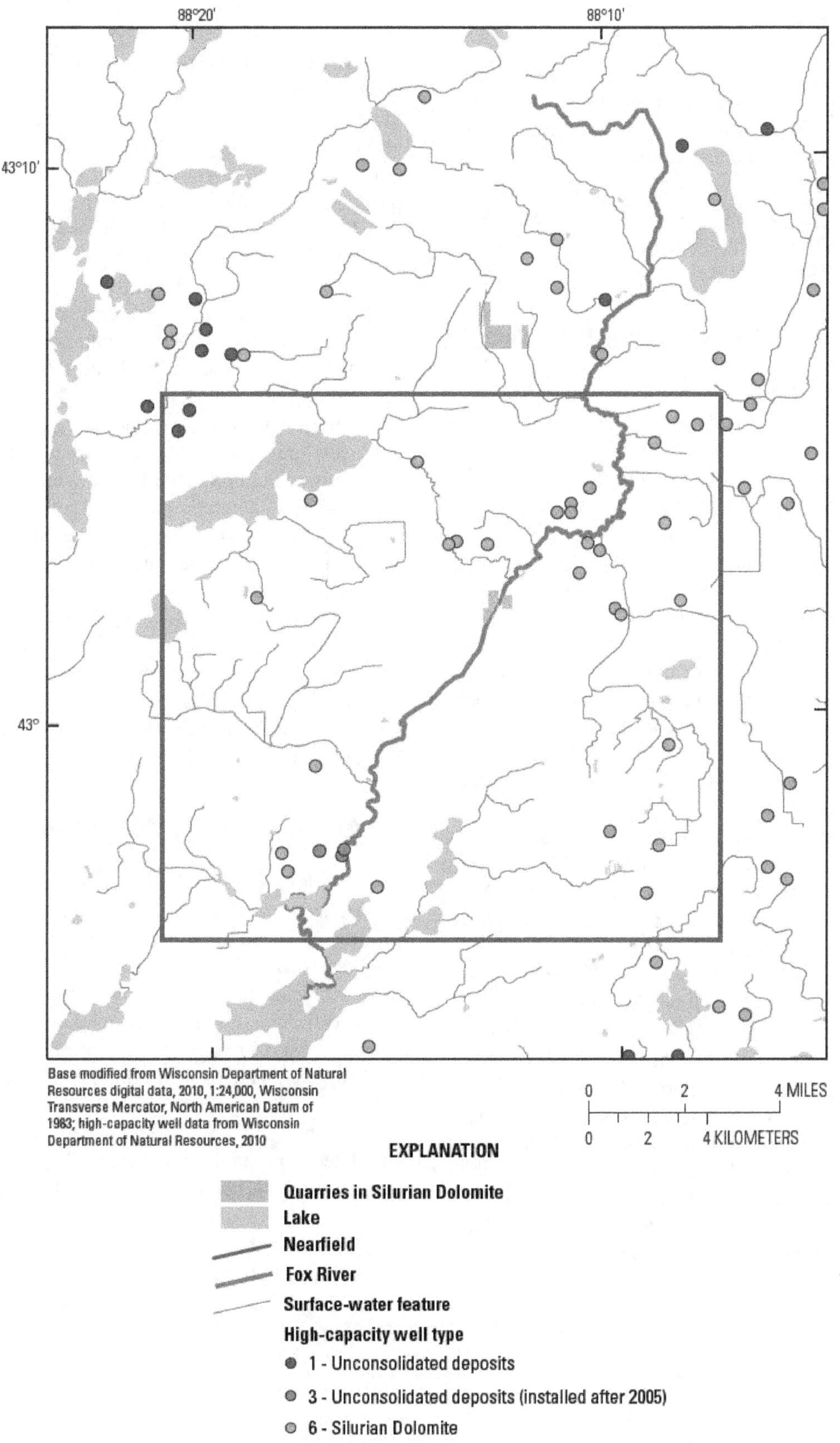

Figure 22. High-capacity wells and quarries in model domain.

4. Model Solver

A new version of MODFLOW-2005 incorporating the Newton formulation of the groundwater-flow equation was recently released by the U.S. Geological Survey (Niswonger and others, 2011). This version is designed to provide a solution for difficult unconfined groundwater-flow problems. In this section, the key features of the computer code are presented and the motives and manner of applying the code in the present study are described.

4.1 MODFLOW-NWT Version of MODFLOW-2005

MODFLOW-NWT is a stand-alone version of MODFLOW-2005 that is intended to overcome instabilities arising from drying and rewetting nonlinearities of the unconfined groundwater-flow equation. It treats nonlinearities of cell drying and rewetting by use of a continuous function of groundwater head, rather than the discrete approach of drying and rewetting used with previous versions of MODFLOW, which resulted in inactive cells when heads fell below the bottom elevation of unconfined cells during any solver iteration. Conversion of cells between active and inactive tends to create discontinuities in calculated flow and can result in convergence failures in many situations. MODFLOW-NWT keeps all active cells that are active at the start of the simulation. It assigns a head to unconfined cells even when the head falls below the cell bottom. Dry cells, although kept active, no longer participate in horizontal aquifer flow.

The Newton formulation calculates intercell conductances in a different way than past versions of MODFLOW in order to maintain continuity in the head field. MODFLOW-NWT uses upstream weighting in the conductance calculation, which allows the formulation of conductance derivatives to remain smooth over the full range of head for a model cell. Instead of converting to zero as a break in a linear slope as the unconfined cell dries, conductance values approach zero along a gradual slope at small saturated thicknesses (fig. 23*A*). A similar smoothing function is enforced for pumping from wells—the discharge does not change abruptly to zero when the head falls below the unconfined cell bottom. At low saturations, the pumping rate is gradually decreased to zero such that the derivatives associated with pumping remain smooth over the full range of heads (fig. 23*B*). The change to the conductance calculation requires use of the Upstream Weighting (UPW) package in MODFLOW-NWT instead of the Block-Centered Flow (BCF), Layer-Property Flow (LPF), or Hydrogeologic-Unit Flow (HUF) packages in MODFLOW-2005. It is worth noting that the input to the UPW package is virtually identical to the LPF package input. One input change of consequence relative to MODFLOW-2005 is an option (in version 1.0.1 of MODFLOW-NWT at the beginning of the WEL package) that allows the user to specify the saturated-thickness threshold at which well discharge is gradually decreased.

The NWT linearization approach generates an asymmetric matrix, which is different from the symmetric matrix generated by the standard Picard formulation of the groundwater-flow equation (Niswonger and others, 2011). The asymmetric matrix cannot be resolved with the solvers accompanying previous versions of MODFLOW. The NWT version contains asymmetric-solver options, including the Orthomin/stabilized conjugate gradient (χMD) solver used in this application. The solution method is made more stable by under-relaxation (a method for calculating the head solution for a particular nonlinear iteration that weights the solution from previous iterations with the present iteration) and, at the discretion of the user, by residual control (a backtracking scheme that slows the solution if the Newtown method overshoots a solution when derivatives change abruptly as a function of head). More solver input is required for the χMD solver relative to packages such as PCG2, but the code allows the user to select a broad set of options with a single choice by specifying the degree of model complexity (simple, moderate, or complex).

4.2 Application of NWT Formulation

The layers in the Upper Fox River Basin models are all defined as unconfined so that the transmissivity of cells is properly adjusted to reflect the degree of saturated thickness. The fine-favored and coarse-favored models share a discretization scheme characterized by thin layers "suspended" from the land surface. For example, the bottom of layer 1 is 20 ft below the land surface; the bottom of layer 2 is 50 ft below land surface. Whenever the simulated water table is more than 20 ft below the land surface, which is expected to occur over much of the model domain, cells in the top layer are dry; simulated water tables more than 50 ft deep (which is expected to occur under hillsides) yield dry cells in both layers 1 and 2.

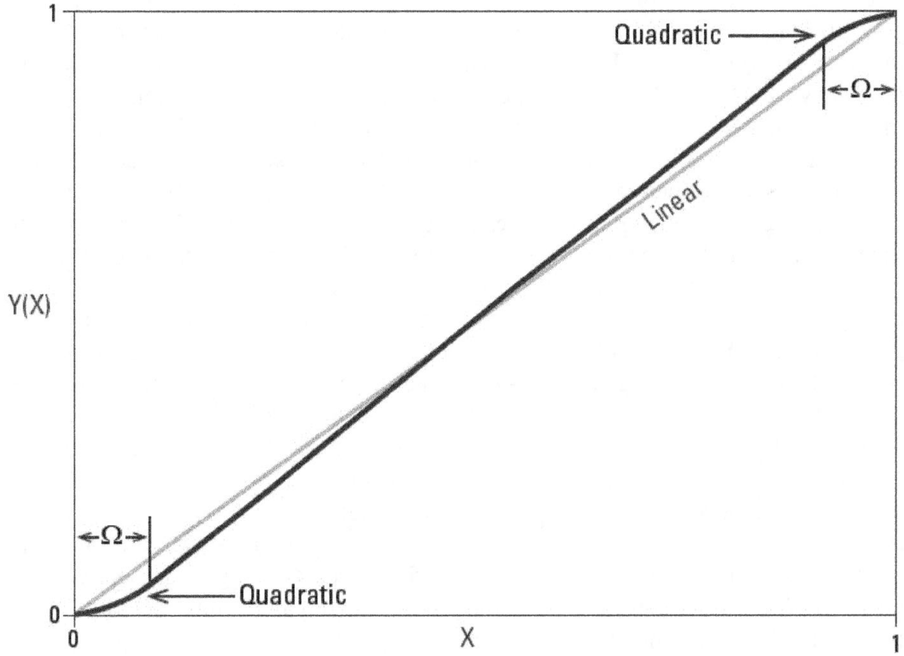

Combined quadratic and linear functions used to smooth conductance and storage in MODFLOW-NWT (black), and a linear function that is used by MODFLOW-2005 (blue). X is the saturated thickness divided by the cell thickness, Y is the value of the smoothing function, and Ω is the interval of X where the quadratic equation is applied, and is equal to 0.1 in this example (NWT input file variable `THICKFACT`).

Figure 23*A*. Smoothing curves used in MODFLOW-NWT formulation—inter-cell conductance and storage (from Niswonger and others, 2011).

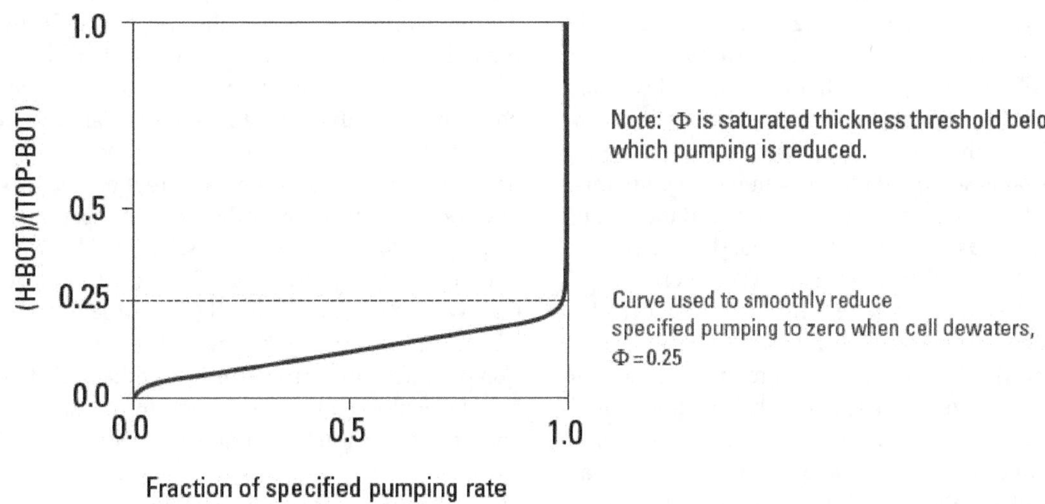

Figure 23*B*. Smoothing curves used in MODFLOW-NWT formulation—pumping (from Niswonger and others, 2011).

The high frequency of dewatered cells creates discontinuities in the head solution and convergence problems with MODFLOW-2005. Initial attempts to resolve the groundwater-flow equation for the Upper Fox models using MODFLOW-2005 and the PCG2 solver resulted not only in many dry cells (which is inevitable given the layer structure), but also severe numerical problems associated with the presence of active cells over inactive cells (fig. 24). For thousands of row/column locations during the solver iterations, layer 1 would remain saturated but layer 2 would become dry, and, therefore, inactive. The layer 1 cells would, in many cases, be isolated from surrounding cells (the intercell conductances falling to zero), and recharge applied to these active cells would accumulate, yielding perversely high water-table elevations. The unacceptable solution shown in the cross section in figure 24 occurred throughout the model domain and was not amenable to stable correction by manipulating input to the PCG2 or other Picard solvers. A partial remedy could be attained by switching more than 5,000 cells from active to inactive status, but the ad hoc changes bias the solution and severely restrict the ways the model can be applied to test, for example, management alternatives involving hypothetical pumping. Attempts to work around the problem, by converting to confined conditions, introduces unacceptable bias in transmissivity assigned to model layers.

The alternative formulation offered by the MODFLOW-NWT program resolves the problem of active over inactive cells and produces a reasonable and smooth water-table surface over the entire model domain as well as reasonable patterns of horizontal and vertical flow "Model Results" section). Virtually all model cells remain active throughout the MODFLOW-NWT solution, although the head in many cells, especially in layer 1, falls below the cell bottom. The only inactive cells in the model are those occupied by Pewaukee Lake because they are handled through the LAK package.

The NWT solver options selected for this study (table 7) correspond to choices associated with each of the three levels of model complexity recognized by the code (for example, see solver input, classified as simple, moderate, or complex, under "Underrelaxation," "Residual Control," and "Linear Solution Options for χMD" in table 7). The head tolerance for achieving a model solution is set to 0.05 ft for both models; the flux tolerance is set to 5,000 ft^3/day for the fine-favored model and to 10,000 ft^3/day for the coarse-favored model (the latter turns out to be somewhat more difficult to solve). These criteria yielded a small overall mass balance error and small local mass balance errors (see "Model Results" section).

The ability of MODFLOW-NWT to produce robust solutions under a variety of steady-state and transient stress conditions for a challenging unconfined problem outweighs several disadvantages from use of the new program. One disadvantage is the number of cells (a total of 1,806,448 active cells in the model domain) requires use of a 64-bit instead of a 32-bit computer architecture to execute the code.

A second disadvantage is that runtimes (averaging about an hour for a steady-state simulation) are longer than for a MODFLOW-2005 run using the PCG2 solver. Part of the increased runtime is because flux convergence is calculated in terms of the root-mean-squared (RMS) flux difference among cells instead of simply the maximum change as it is for PCG2. The RMS criteria was necessary in MODFLOW-NWT because it is required for residual control (backtracking). RMS can be a better criteria for convergence because it is an indicator of the residuals of all cells rather than the error for the cell with the largest residual. For example, if all the cells have an error slightly less than the largest error, then this solution will have a larger RMS than the case where nearly all the cells have a low residual and one or a few have larger residuals. For the Upper Fox simulations, the RMS criterion was always more strict than the maximum flux difference and, therefore, required more iterations.

A third issue is less a disadvantage than a particular feature of the solution arising from the Newton linearization method. As discussed above, the MODFLOW-NWT formulation reduces the well discharge when the saturated thickness in a cell hosting a pumping well falls below a prescribed saturation threshold. In the Upper Fox River Basin models, the saturation threshold is set to the default value of 20 percent of the total layer thickness in MODFLOW-NWT (for example, in the case of a well pumping from layer 1, which normally has a thickness of 20 ft unless the unconsolidated sequence is very thin, discharge begins to be reduced when the water table falls more than 16 ft below land surface, the top of layer 1). As in the case of the Picard formulation and MODFLOW-2005, the loss of pumping is a function of the head solution, which in turn is a function of parameter input. Unlike MODFLOW-2005, the well discharge supported by the solution is not at its full rate or at zero but can attain any rate at or between the full rate and zero as a function of simulated saturated thickness. The summary MODFLOW-NWT output LST) file tabulates the simulated discharge in wells that experience reduction from the desired input rate. This new paradigm adds a complication to the interpretation of the output, but it has the compensating advantage that the simulated withdrawals from wells under unconfined condition reflect the capacity of the unconfined aquifers to support target or prescribed discharge rates on the basis of the values of input parameters. For the Upper Fox River Basin models, the NWT code does reduce pumping from reported rates for some production wells, but the overall reduction is small (see "Model Results" section). Reduction in pumping because of cell drying can be an indication of calibration error (neglecting the uncertainty in reported rates) because the simulation aquifer is unable to provide the amount of water that the real aquifer provides.

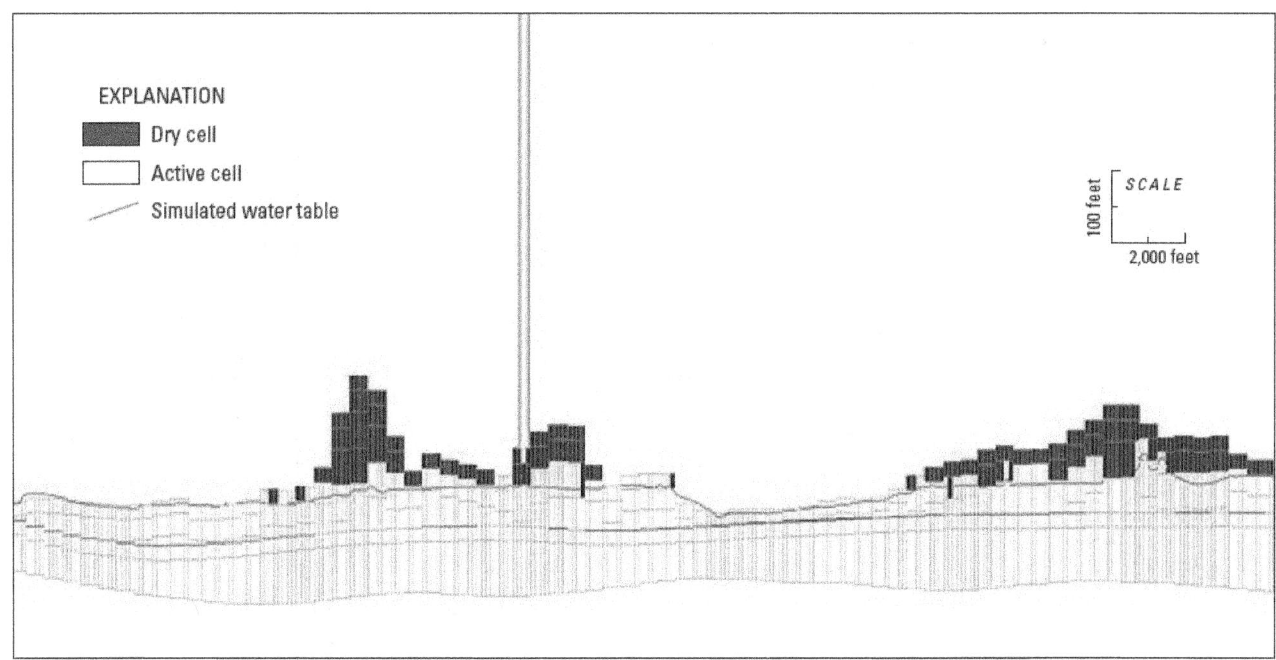

Figure 24*A*. Simulated water-table configuration using MODFLOW-2005 and MODFLOW-NWT—MODFLOW-2005 with PCG2 solver in cross section. Example results in cross section are for calibrated coarse-favored inputs along Row 298 (see fig. 7 for section trace).

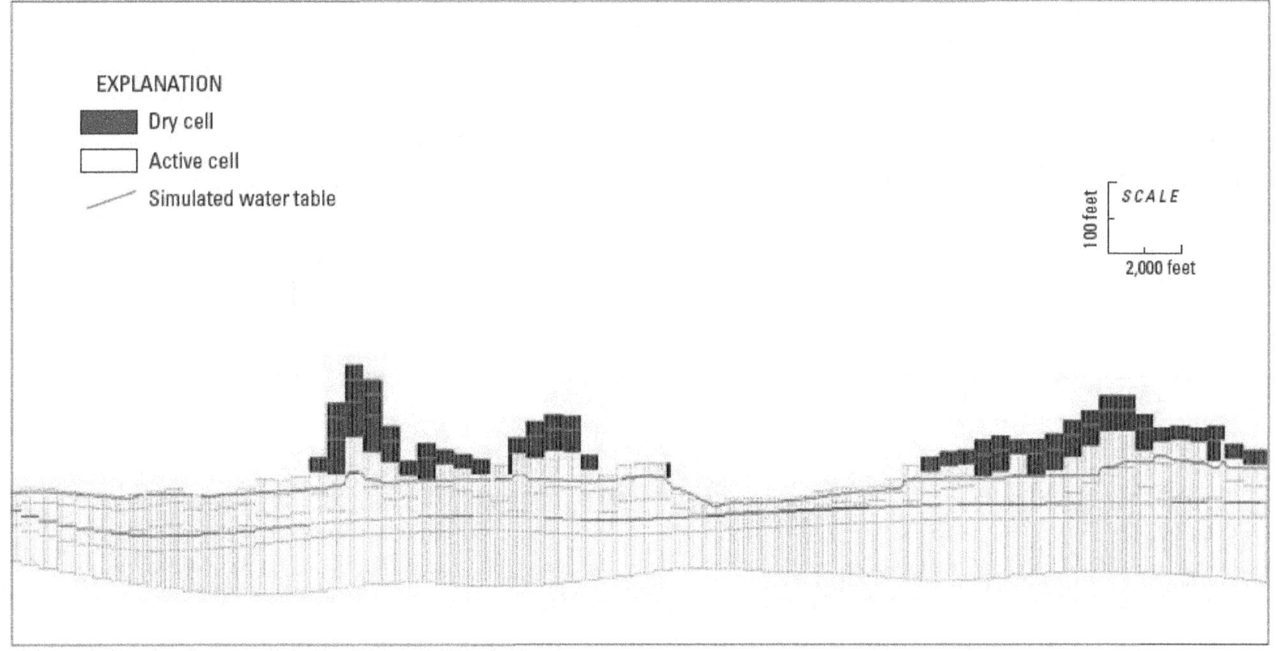

Figure 24*B*. Simulated water-table configuration using MODFLOW-2005 and MODFLOW-NWT—MODFLOW-NWT with XMD solver in cross section. Example results in cross section are for calibrated coarse-favored inputs along Row 298 (see fig. 7 for section trace).

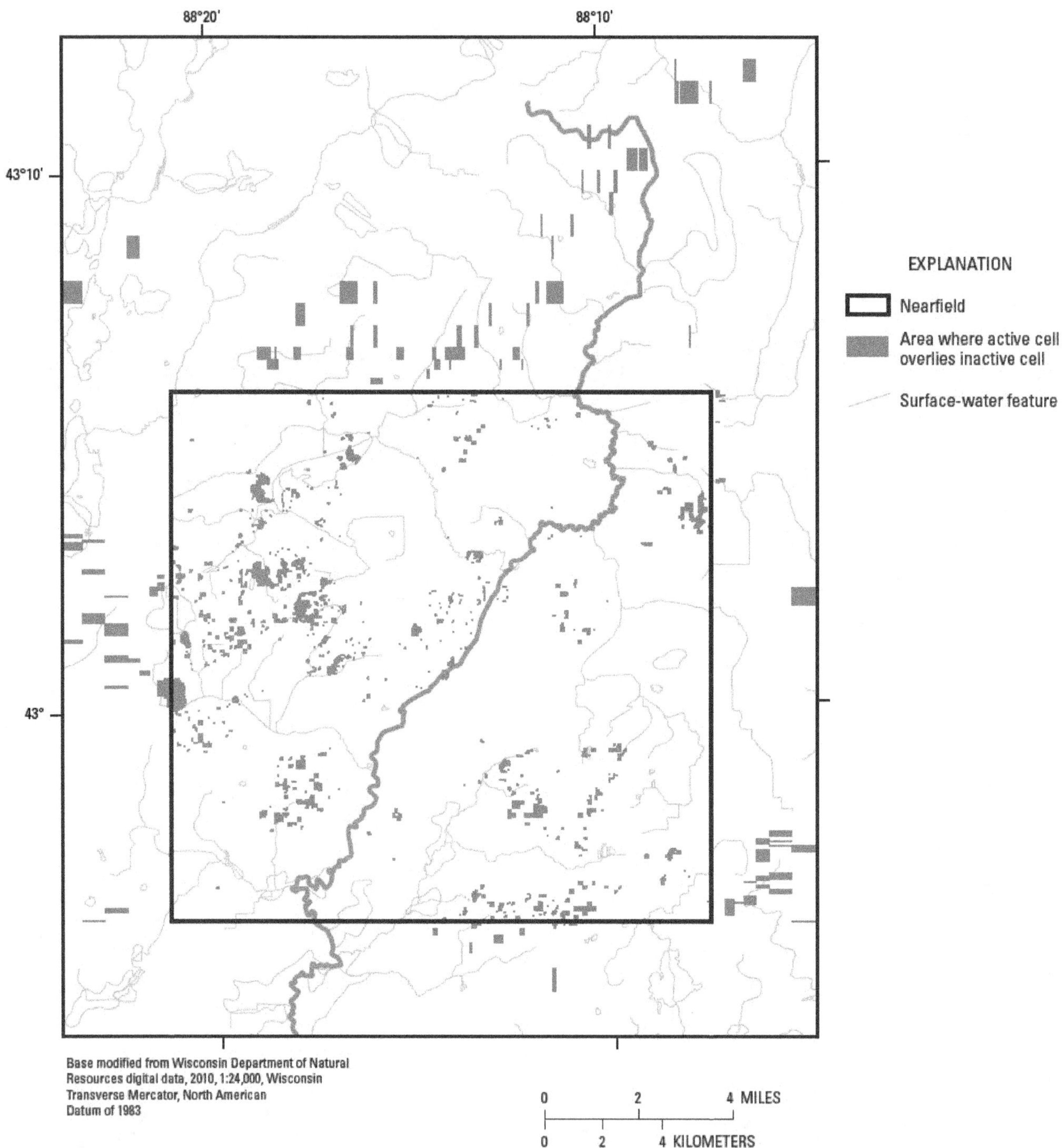

Figure 24*C*. Simulated water-table configuration using MODFLOW-2005 and MODFLOW-NWT—MODFLOW-2005 with PCG2 solver in plan view. Red cells are locations where active cells overlie inactive cells for calibrated coarse-favored model, producing excessively high simulated water-table elevations (see fig. 24a). (Total number of cell locations with excessively high simulated water-table elevations is 6,578.)

Table 7. Suggested and selected input values for the NWT input file

[m, meter; ft, foot; m^3/d, cubic meters per day; ft^3/d, cubic feet per day; --, indicates that values are not necessarily applicable; selected values are in **bold-face** type. For explanation of variables, see Niswonger and others, 2011, section titled "NWT input file"]

Input variable name	Default values	Selected values		
		Iteration control		
HEADTOL	*1x10^{-4} (m)	**0.05 ft**		
FLUXTOL	*500 (m^3/d)	**5,000 ft^3/d for fine-favored; 10,000 ft^3/d for coarse-favored**		
MAXITEROUT	100	**300 for fine-favored; 1,000 for coarse-favored**		
		Dry-cell tolerance		
THICKFACT	0.00001	**0.00001**		
		NWT options		
LINMETH	1	**2** → χMD linear solution control		
IPRNWT	0	**1** → print max head change		
IBOTAV	**0	**1** → corrections will be made to groundwater head relative to the cell-bottom altitude if the cell is surrounded by dewatered cells		
		Underrelaxation input		
		Model complexity		
	Simple	Moderate	Complex	
DBDTHETA	0.97	0.7	0.4	**0.5**
DBDKAPPA	.0001	.0001	.00001	**.00001**
DBDGAMMA	.0	.0	.0	**.0**
MOMFACT	.0	.1	.1	**.0**
		Residual control		
		Model complexity		
	Simple	Moderate	Complex	
BACKFLAG	***0	0	0	**1** → Residual control active
MAXBACKITER	--	--	--	**20**
BACKTOL	--	--	--	**1.05**
BACKREDUCE	--	--	--	**.6**
		Linear solution control and options for GMRES – **not used**		

Table 7. Suggested and selected input values for the NWT input file.—Continued

[m, meter; ft, foot; m^3/d, cubic meters per day; ft^3/d, cubic feet per day; --, indicates that values are not necessarily applicable; selected values are in **bold-face** type. For explanation of variables, see Niswonger and others, 2011, section titled "NWT input file"]

Input variable name	Default values			Selected values
Linear solution control and options for χMD – **used**				
	Model complexity			
	Simple	Moderate	Complex	
IACL	2	2	2	**2**
NORDER	1	1	0	**0**
LEVEL	0	1	3	**8**
NORTH	2	2	7	**2**
IREDSYS	0	0	0	**0**
RRCTOLS	.0	.0	.0	**.0**
IDROPTOL	1	1	1	1
EPSRN	$1x10^{-3}$	$1x10^{-3}$	$1x10^{-4}$	**$1x10^{-3}$**
HCLOSEχMD	$1x10^{-4}$	$1x10^{-4}$	$1x10^{-4}$	**$1x10^{-3}$ ft**
MAXITINNER	50	50	50	**100**

*These values are dependent upon the units specified in the MODFLOW-2005 discretization input file. Values given are for units of meters and days.

**The optimal value for IBOTAV is problem specific. Values of 0 and 1 should be tested for each problem.

***BACKFLAG should be set to 0 (residual control set to inactive) unless there are convergence problems. "Options" must be set to "Specified" if the residual control option is used.

The implementation of the NWT solution method to the Upper Fox River Basin models was subjected to three sensitivity tests:

- An alternative solver algorithm, called GMRES, was substituted for χMD;
- Residual control was deactivated; and
- Starting heads equal to the land-surface elevation at each row/column location were substituted for starting heads corresponding to the preliminary solution with initial parameter values (for which the average water-table depth from land surface is on the order of 35 ft).

A variety of metrics—including head distribution, average depth to the water table, number of dry cells, and water budget terms—for comparing the base results for the fine-favored and coarse-favored models to the solutions for the three sensitivity simulations show negligible difference in output. The major difference among the simulations was runtime—the GMRES solver required more execution time than the XMD solver and, not surprisingly, a solution starting from heads equivalent to the land surface required more execution time than a solution starting from heads corresponding to preliminary output. At least for the sensitivity tests conducted, the application of the NWT formulation is robust.

The version of the MODFLOW-NWT program used in this application is not the final published code (which was not released until the modeling was completed), but a development code made available to USGS researchers several months before the May 2011 release. The development code differs from the final version in two ways:

- Several packages (UPW, SFR, NWT and WEL) have slightly different input formats although the options in the development and published versions are identical; and
- The LAK package algorithm is updated in the final version, partly to reduce simulation time.

As a check on the stability of the code, the fine-favored model calibrated with the development version also was run with the published version of MODFLOW-NWT. The results of the published version are very similar to those of the development version. For example, in comparing the global water budgets for the simulations, the biggest difference occurs for the net flow to streams represented by the SFR package, and the difference is less than 0.04 percent.

Because the development version of the program was used to generate the results discussed in this report, it is included in the model archive. The source FORTRAN files and compiled executable correspond to the version of the code available on January 7, 2011.

5. Model Calibration

The calibration process involves updating the initial estimates for selected input values in order to improve the match between simulated groundwater levels and flows and corresponding field observations called *targets*. Several techniques are employed to enhance the calibration process and make it tractable for *parameter estimation* (that is, estimate of independent variables—parameters—through observed values of dependent variables—head and flow targets). The techniques include selection of parameters subject to estimation; assembly of diverse target types to better explore the parameter space; setting of target weights; and, after preliminary adjustment of parameters by manual trial-and-error, application of nonlinear regression by means of singular value decomposition with regularization. For this study, the nonlinear regression is implemented by use of the code PEST (Doherty, 2008a, 2008b). In the end, the calibration process not only improves the fit between measured and simulated targets, but also adds insight into the major controls on the flow system through identification of parameters to which model results are most sensitive. A key element of the calibration process is the application of the parameter estimation methods to the fine-favored and coarse-favored models.

5.1 Estimated Parameters

Several sets of model input parameters are estimated as part of the calibration process. As discussed in the "Model Construction" section, the unconsolidated material in model layers 1 through 5 is mapped to five facies, with distinct facies zones for the fine-favored and coarse-favored models (fig. 15). The K_h assigned each facies constitutes one set of five estimated parameter values; the K_v constitutes a second set of five values. In the calibration process, only the K_v value of the zone corresponding to the dominantly fine facies is directly estimated. The values for the other zones are estimated as ratios relative to the neighboring finer zone. A third set of K_v values is associated with the three streambed zones applied in the Upper Fox River Basin (fig. 20). A fourth set of inputs corresponds to the K_h and K_v of the Silurian dolomite in model layers 6 and 7. A single multiplier is applied to all the K_h and to all the K_v zones of the dolomite copied from the SEWRPC regional model. In effect, the calibration process is used to improve the estimate of the bedrock unit's overall transmissivity and vertical resistance. The last parameter set estimated is particular to the use of the Upper Fox River Basin models in transient mode—that is, the storage properties of the unconsolidated deposits in response to a time-dependent stress—expressed as uniform values for specific storage and specific yield. One parameter set that is not subject to the calibration process is the zonation of recharge (fig. 17). The zonal values were estimated by a method dependent on matching recharge to base flow in tributary basins (see "Model Construction" section under "Water Withdrawals"). As a result only limited information is gained by reestimating these values on the basis of base-flow targets applied to the model calibration process and discussed below. One motive for fixing the recharge values in the models is to simplify the calibration process and make it more sensitive to the uncertainty of the various hydraulic conductivity values. A second motive for fixing the input is to insure that the recharge values are consistent with the flux boundary conditions derived from the regional SEWRPC model and applied at the lateral edges and bottom of the Upper Fox River Basin model.

A listing of estimated parameters along with initial values can be found in table 8. It is important to emphasize that the parameters estimated and the initial values assigned to them are the <u>same</u> for the fine-favored and coarse-favored models. However, the calibration process ultimately yields distinct parameter values for the two models.

5.2 Calibration Targets

The Upper Fox River Basin models are calibrated to five target types. In the estimation process, different parameter sets are expected to be sensitive to different target types. The target values themselves are all affected by error and uncertainty.

5.2.1 Water Levels

Water levels simulated by the model are compared to a map of the generalized water table in southeastern Wisconsin prepared by researchers at the Wisconsin Geological and Natural History Survey (Southeastern Wisconsin Regional Planning Commission and Wisconsin Geological and Natural History Survey, 2002, Map 21). Calibration targets are obtained by sampling the published contours at a 2,000-ft interval within the model domain. This process yields 3,492 water-level targets: 2,099 in the farfield and 1,393 in the nearfield (fig. 25). The median depth of the water levels below land surface at the target locations is 36 ft. Taking the targets as approximations of the water-table elevation implies that layer 1, bottoming 20 ft below land surface, will be dewatered over much of the model domain.

Table 8. Parameters estimated in calibration process.

[ft/d, foot per day; kx, horizontal hydraulic conductivity; kv, vertical hydraulic conductivity; ksfr, streambed hydraulic conductivity]

Parameter	Initial value	Lower bound	Upper bound
Horizontal hydraulic conductivity (ft/d)			
kx2 (dominantly fine)	0.50	0.10	1.00
kx4 (relatively fine)	2.00	1.00	4.00
kx12 (mixed fine and coarse)	10.00	4.00	20.00
kx13 (relatively coarse)	40.00	20.00	60.00
kx14 (dominantly coarse)	80.00	60.00	200.00
Vertical hydraulic conductivity (ft/d)			
kv2 (dominantly fine)	.01	.002	.02
Ratio kv4 to kv2 (for relatively fine)	2.00	1.00	100.00
Ratio kv12 to kv4 (for mixed fine and coarse)	2.00	1.00	100.00
Ratio kv13 to kv12 (for relatively coarse)	10.00	4.00	100.00
Ratio kv14 to kv13 (for dominantly coarse)	10.00	4.00	100.00
Streambed hydraulic conductivity (ft/d)			
ksfr_low	1.00	.10	5.00
ksfr_middle	5.00	.50	25.00
ksfr_high	25.00	2.50	100.00
Multiplier on dolomite hydraulic conductivity zones			
kx	1.00	.02	4.00
kv	1.00	.02	4.00
Multiplier on storage values of unconsolidated deposits			
specific storage (ft^{-1})	1.00	.10	10.00
specific yield (--)	1.00	.10	10.00
Multiplier on recharge zones (fixed)			
recharge	1.00	1.00	1.00

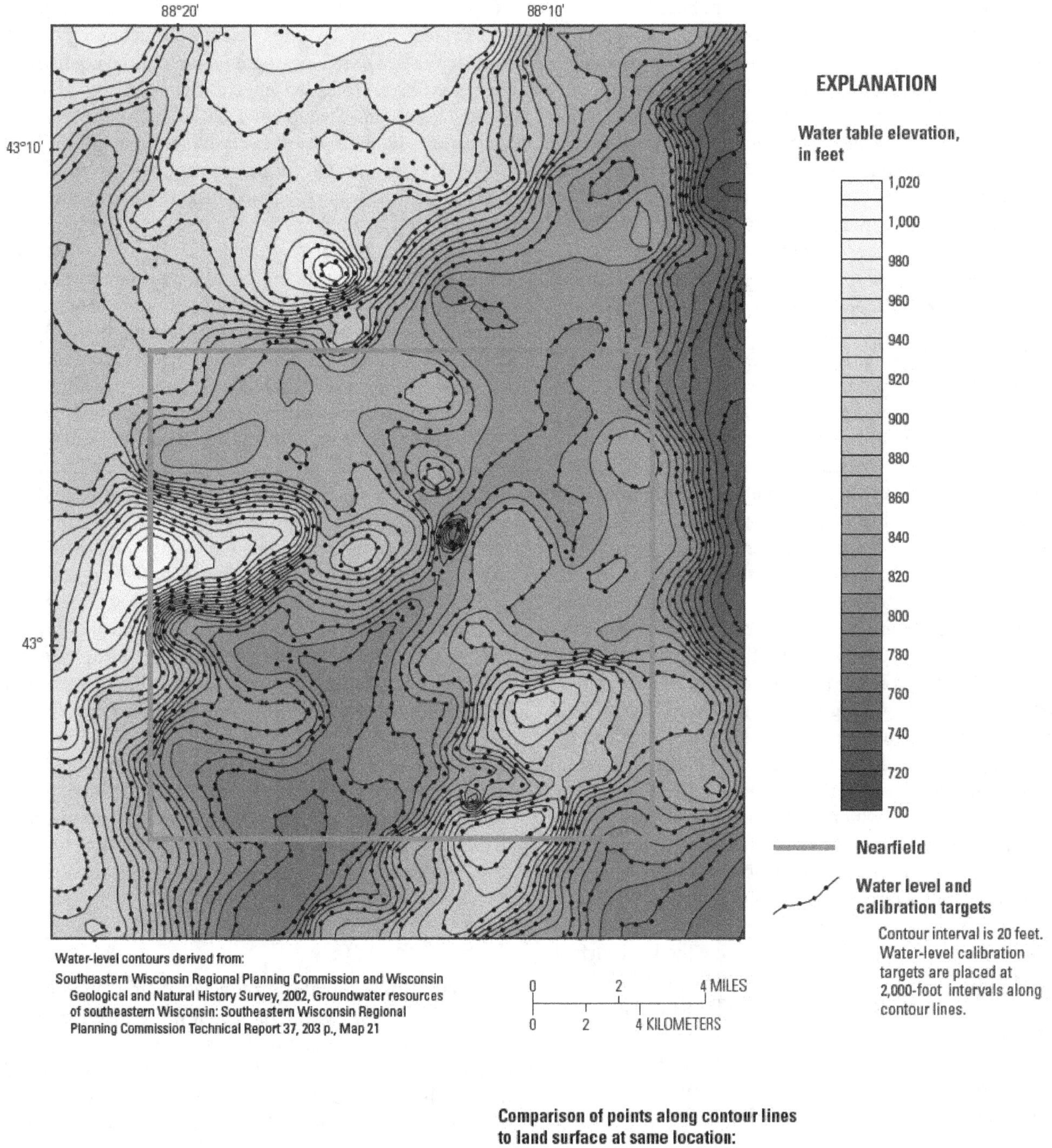

Comparison of points along contour lines to land surface at same location:

Median depth .... 36.1 feet

Average depth .... 41.6 feet

Standard deviation depth ...... 30.9 feet

Minimum depth .... 0.1 foot

Maximum depth .... 213.3 feet

Figure 25. Calibration targets: water levels.

The water-level contours in the map are based chiefly on static water levels recorded by drillers and geologists in wells open to the shallow aquifer, on elevations corresponding to groundwater outcrops at surface-water bodies, and on topography. The 2002 Wisconsin and Natural History Survey report states that "it was beyond the scope of this study to field-check the locations and water level measurements reported on DNR well constructor's reports that were used to construct the water-table map. The map does not reflect local details in the water because the contour lines were generalized and interpolated from the nearest data points, and natural fluctuations in the water table can be as much as 15 feet." Another, although smaller, source of uncertainty arises from the fact that the static water levels do not correspond exactly to the water table itself, but to the elevations below the water table where the wells are open. Given the presence of even small downward or upward vertical gradients within the shallow aquifer system, the implied water-table surface will fall below or above the recorded water levels. (For example, if the average vertical gradient is 0.01 ft/ft downward and the well is open 100 ft below the water table, then the recorded water level is 1 ft below the implied water table). Inspection of the well database used to construct the map (Peter Schoephoester, Wisconsin Geological and Natural History Survey, written commun., January 12, 2010) indicates that the median depth of the 1,679 shallow wells listed is on the order of 92 ft and that the median depth of the casing interval is on the order of 89 ft (water wells typically have short screens), with 80 percent of the casing depths between 42 ft and 200 ft. On the basis of this information, the calibration targets were assigned to unconsolidated model layer 4, which when active is 100 to 150 ft below land surface but when pinched is less than 100 ft below land surface. To account for the influence of surface-water elevations and topography on mapped contours, especially in low-lying areas, the water-level targets were assigned to model layer 2 instead of model layer 4 where the land surface is below 860 ft, an area corresponding to the Upper Fox River Valley and confluent valleys (see fig. 5*B*). These layer adjustments are made to better align the contour-derived targets with the underlying water-level data, but it is recognized there is uncertainty around target values on the order of as much as 15 ft because of natural fluctuations in head.

5.2.2 Vertical Head Differences

The estimation of K_v in the calibration process is greatly enhanced by the availability of targets linked to the vertical head difference between layers. Public water-supply projects for the city of Waukesha included the installation of test borings at depth and the collection of water levels measured relative to land surface. In particular, data collected as part of studies conducted near a wetland between the Fox River and Pebble Brook (Aquifer Science & Technology, 2010) and along the Fox River upgradient from the confluence of Genesee Creek (John Jansen, Aquifer Science & Technology, written commun., December 1, 2010) allow the estimation of vertical gradient targets at the two locations (fig. 26).

The open interval for the test borings was around 100 ft depth. Depth to water was measured relative to the top of casing, noted to be about 2.5 ft above land surface. Six of the eight test borings registered water levels above land surface of 1 to 3 ft, implying "flowing conditions" and upward gradients with respect to the water table located below land surface. Only test borings labeled as "lath1" and "lath10" (fig. 26) registered water levels below land surface. No direct measurements of the water-table elevation were available at the test boring locations because they are typically open 100 ft below land surface and tens of feet below the water table. In order to construct a vertical head difference target, the depth of the water table was assumed to be 1 ft below land surface at all the test boring locations except for "lath1," because it is farther from surface water, and the water table was assumed to be 3 ft below land surface. The observed target corresponds to the difference between the assumed water table and the measured water level in the test borings; the simulated target corresponds to the difference between the head in layer 1 and layer 4. The uncertainty of the targets is high given all the assumptions incorporated in estimating the observed values, but one qualitative test of the model fit is its ability to reproduce "flowing conditions" at the appropriate locations. The simulation of strong upward gradients and "flowing conditions" depends not only on the proximity of the test borings to surface-water features but also on the vertical resistance to flow imposed by the K_v assigned the unconsolidated facies.

5.2.3 Base Flow

Base-flow estimates offer a calibration target linked to groundwater flow instead of groundwater levels. Even though the main source of water to the groundwater system, the recharge input, is fixed for the fine-favored and coarse-favored models, base-flow targets still afford a way to check the model's ability at specific locations to properly partition flow between circulation to shallow discharge zones represented by surface water as opposed to deeper movement toward wells and model boundaries.

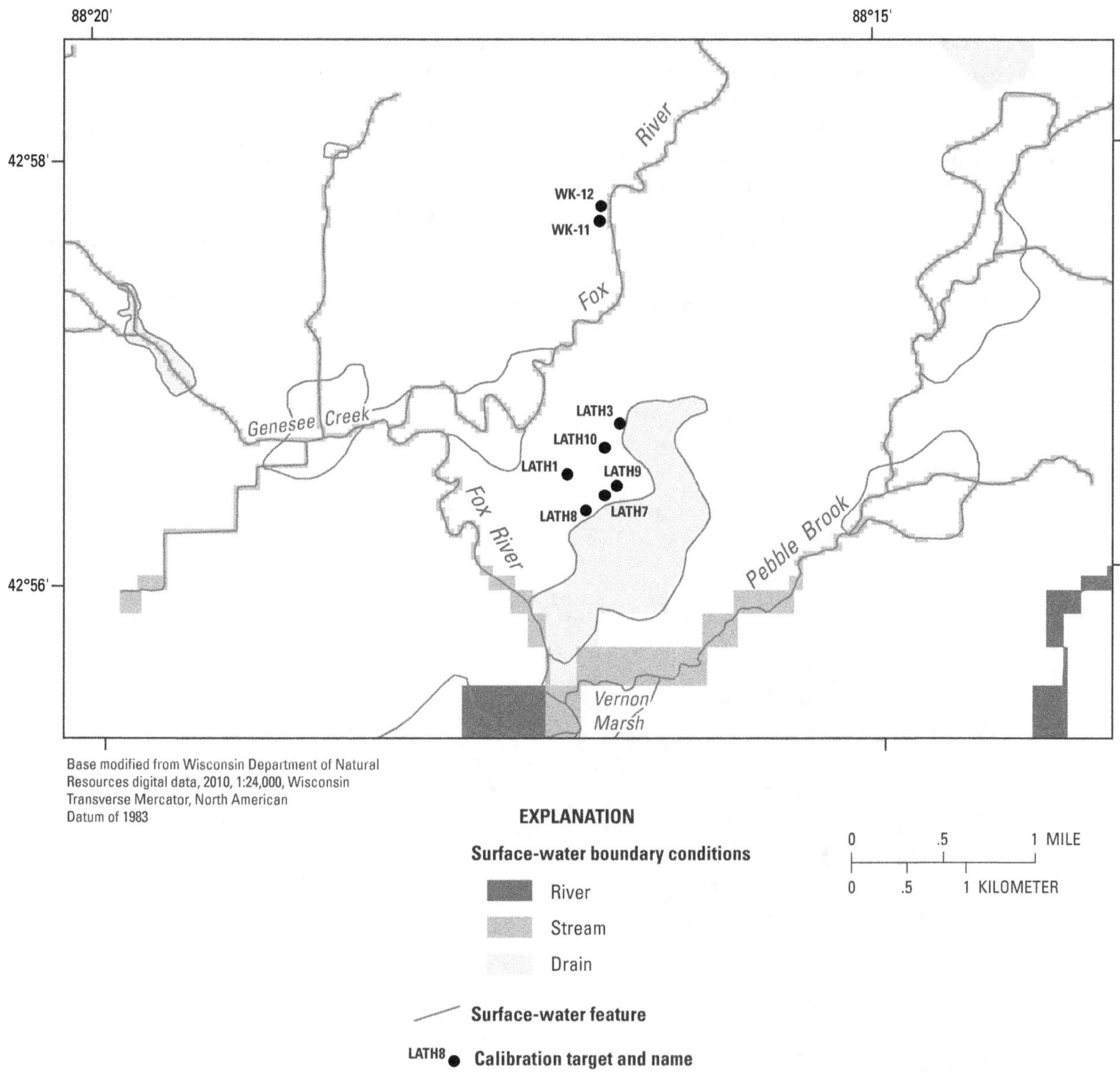

Figure 26. Calibration targets: vertical head difference.

Streamflow records are used to estimate base flow U.S. Geological Survey, 2010c). Records from two USGS streamgages are available along the Upper Fox River (fig. 6*B*)—the Waukesha gage for the period January 1, 1963, to September 30, 2008, and the Watertown gage for the period December 1, 1992, to September 30, 2000 (this gage was discontinued after 2000). It is useful to relate base flow to the flow-duration curve, a plot that shows the percentage of time that flow in a stream is likely to equal or exceed some specified value (accordingly, the 50-percent value (Q_{50}) defines the flow exceeded 50 percent of the time, the 75-percent value (Q_{75}) defines the flow exceeded 75 percent of the time, and the 90-percent value (Q_{90}) defines the flow exceeded 90 percent of the time). A method developed for Wisconsin by Gebert and others (2007) estimates annual base flow for the basin upgradient from a gage in terms of basin area and a base-flow factor proportional to the 90-percent flow-duration value according to the equation:

$$Q_b = 0.906 A^{1.02} B_f^{0.52} \tag{3}$$

where

- Q_b is average annual base flow, in cubic feet per second,
- A is drainage area, in square miles, and
- B_f is the base-flow factor, defined as the 90-percent flow-duration value ($Q_{90,}$ in cubic feet per second) divided by the drainage area, in square miles.

For this statistically significant regression (p is less than 0.001), the statewide average standard error of estimate was 12 percent, and the statewide R^2 (a measure of how well a regression line approximates the data points, varying between 0 and 1, with 1 representing a perfect fit) was 0.992. Application of the equation to the Waukesha gage yields a base-flow estimate of 49.6 ft^3/s. This value is typical of total streamflow in the low-flow period between July and November (fig. 27), although, as expected, it is only a fraction of the typical streamflow in other months when overland flow from precipitation and snowmelt dominates the base-flow component. The average base flow calculated from the Gebert formula is intended to reflect long-term conditions subject to average recharge over the basin rather than the actual 2005 low-flow conditions at the Waukesha gage, which averaged 33.7 ft^3/s between July and November 2005, reflecting precipitation patterns in the spring and summer of 2005.

Application of the equation to the Watertown gage was biased by the short period of record. To overcome the bias, the Q_{90} for the Watertown and Waukesha gages were calculated for the 1992–2000 period over which the Watertown gage was active, and then the ratio of these values was multiplied by the Q_{90} for the full period of the Waukesha gage to generate a corresponding full period Q_{90} for the Watertown gage. This last value then was applied to the base-flow factor in the regression equation. The resulting base-flow estimate for the Watertown gage is 33.6 ft^3/s. The implied base-flow gain from the Watertown to the Waukesha location (including base flow from Pewaukee Lake and Pewaukee River) is about 16 ft^3/s.

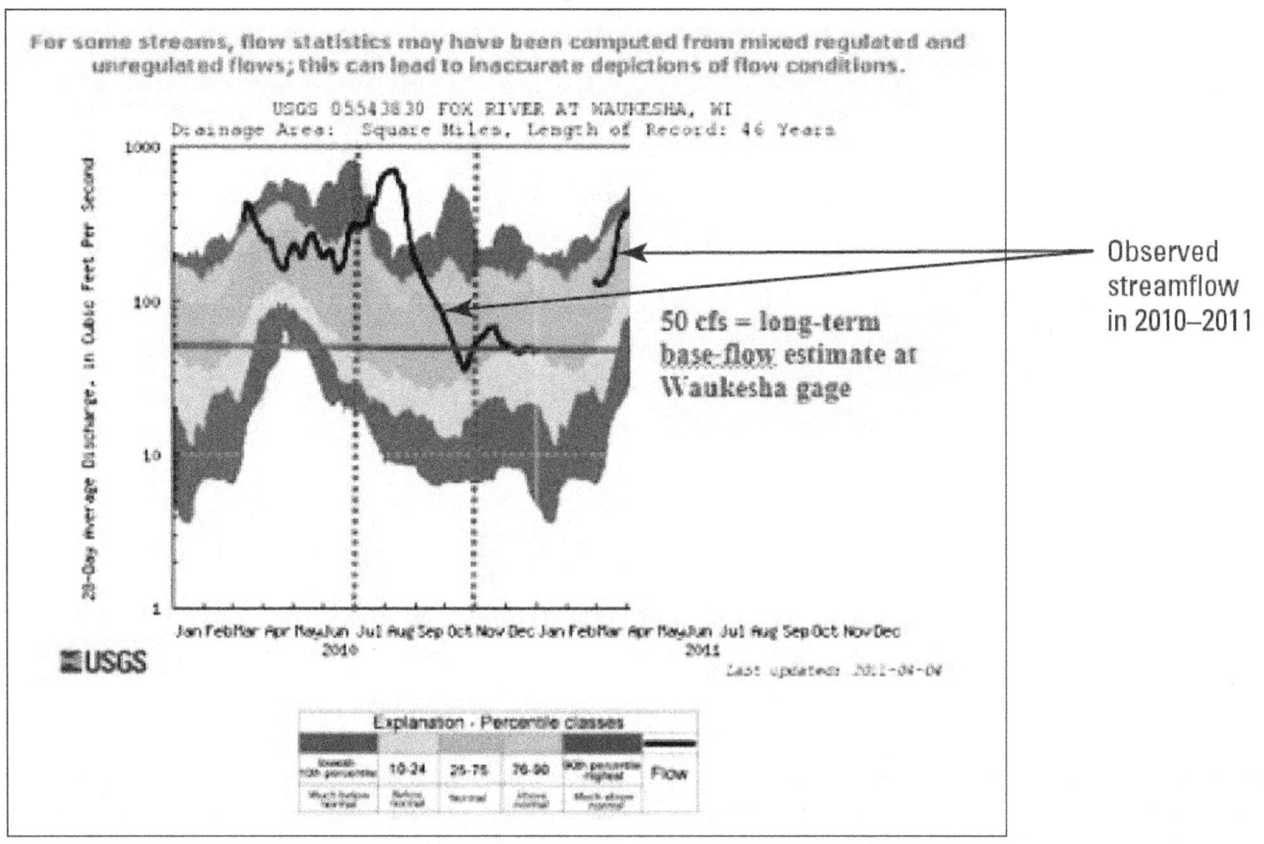

Plot shows flows at Waukesha gage 05543830 from January 2010 to April 2011 superimposed on long-term streamflow pattern.

Image downloaded on April 4, 2011, from:
http://waterwatch.usgs.gov/new/index.php?sno=Waukesha&yr=2011&dt=dv28d&btnGo=GO&m=sitedur&s=&sno2

Figure 27. Comparison of base-flow estimate to streamflow record at Waukesha gage.

The base-flow estimates for the gages are derived from records that include not only groundwater discharge to surface water but also the addition of effluent at two WWTPs upgradient from the target locations of Sussex and Brookfield (fig. 6*B*). The combined average inflow from these WWTPs, corresponding to rates for 2008–9, is 13.29 ft^3/s (see "Model Construction" section under "Input to MODFLOW-2005 Surface-Water Packages"), equal to 40 percent of the base-flow estimate at Watertown and 27 percent of the estimate at Waukesha. However, the effluent flux was probably smaller than the 2008–9 rates for most of the 1963–2008 period of streamflow record because the area developed gradually, opening the possibility that the model will tend to oversimulate observed base flow for 1963–2008 by adding recent rates of effluent discharge to simulated groundwater discharge. Conversely, development over time led to more pumping, implying that the model will tend to undersimulate average base flow by applying 2005 rates of well discharge. These biases add to the uncertainty inherent in the base-flow estimation by regression.

5.2.4 Land-Surface Constraints

The surface-water network input to the model does not contain all groundwater-discharge areas active in the domain. Most notably, groundwater tends to discharge along waterway corridors at various locations that are difficult to map. It is convenient to conceptualize this outflow as part of base flow because of its tendency to exit the groundwater system and immediately discharge as runoff to streams. In the absence of explicit discharge zones inserted at many locations in the model along surface-water bodies, this discharge will be simulated as base flow to the adjacent stream rather than as outflow to the adjacent land surface, but often these riparian locations are marked by water-table conditions simulated above land surface. In other words, "groundwater flooding" in riparian cells is an expected model outcome, which ordinarily represents an acceptable simplification of reality. In contrast, simulated flooding elsewhere in areas of higher topographic elevation is likely to indicate areas of poor model fit. For this reason, two land-surface constraints have been added to the calibration targets:

- a target to minimize average flooding in the group of model cells where water-table elevations are simulated above land surface; and
- a target to minimize the percent of *nearfield* model cells reporting simulated water-table elevations more than 3 ft above land surface.

These targets are constructed in a way that tends to allow riparian groundwater flooding but discourages it elsewhere in the model domain.

5.2.5 Aquifer Test Drawdown

The water-level, vertical-head difference, base flow, and land-surface targets are all applied to the steady-state model intended to approximately reproduce 2005 base-flow conditions. A last target type refers to the drawdown response to a transient stress in the form of an aquifer test conducted on a candidate public supply well installed in the unconsolidated material about one-half mile west of the Fox River (fig. 28). The aquifer test occurred in November 2007 at the site of the future well WK-13 (which began operation in 2009), about 2 years after withdrawals began from nearby consolidated wells WK-11 and WK-12 at a combined average rate of 1,235 Mgal/d over 2006–7 (Jeff Detro, Waukesha Water Utility, written commun., January 18, 2010).

The withdrawal at WK-13 was superimposed on the 2006–7 pumping from wells WK-11 and WK-12, which in turn was superimposed on 2005 conditions. In order to simulate the aquifer-test response to pumping, two transient stress periods were added to the base steady-state simulation, the first of 2-years duration and the second corresponding to the drawdown phase of the aquifer test. Simulated drawdown during the second transient stress period is calculated relative to the water levels at the end of the preceding 2-year stress period. One aquifer test, conducted by Aquifer Science and Recovery (2008), involved 20 hours of pumping WK-13 at 600 gallons per minute (gal/min) and recording of water-level changes at several test borings acting as observation wells, including ENGL-1 and ENGL-5 located 70 and 280 ft, respectively, from WK-13 (fig 28*A*). Shallow well OW-1, located near WK-13 but screened at a depth corresponding approximately to model layer 1 (0 to 20 ft below land surface), also was monitored during the test but showed no detectable response (John Jansen, formerly Aquifer Science & Technology, oral commun., February 2010). Stratigraphic and well-construction records indicate that WK-13 drew water from an open interval corresponding to sandy deposits present from about 85 to 105 ft below land surface (fig. 28*B*). This depth intersects the lower part of model layer 3 (extending 50 to 100 ft below land surface) and the thin vertical extent of model layer 4 (extending 100 ft to 105 ft below land surface at this location). In the fine-favored and coarse-favored models, the layer 3 cell at the WK-13 location is identified with the mixed facies and layer 4 with the coarse-dominated facies.

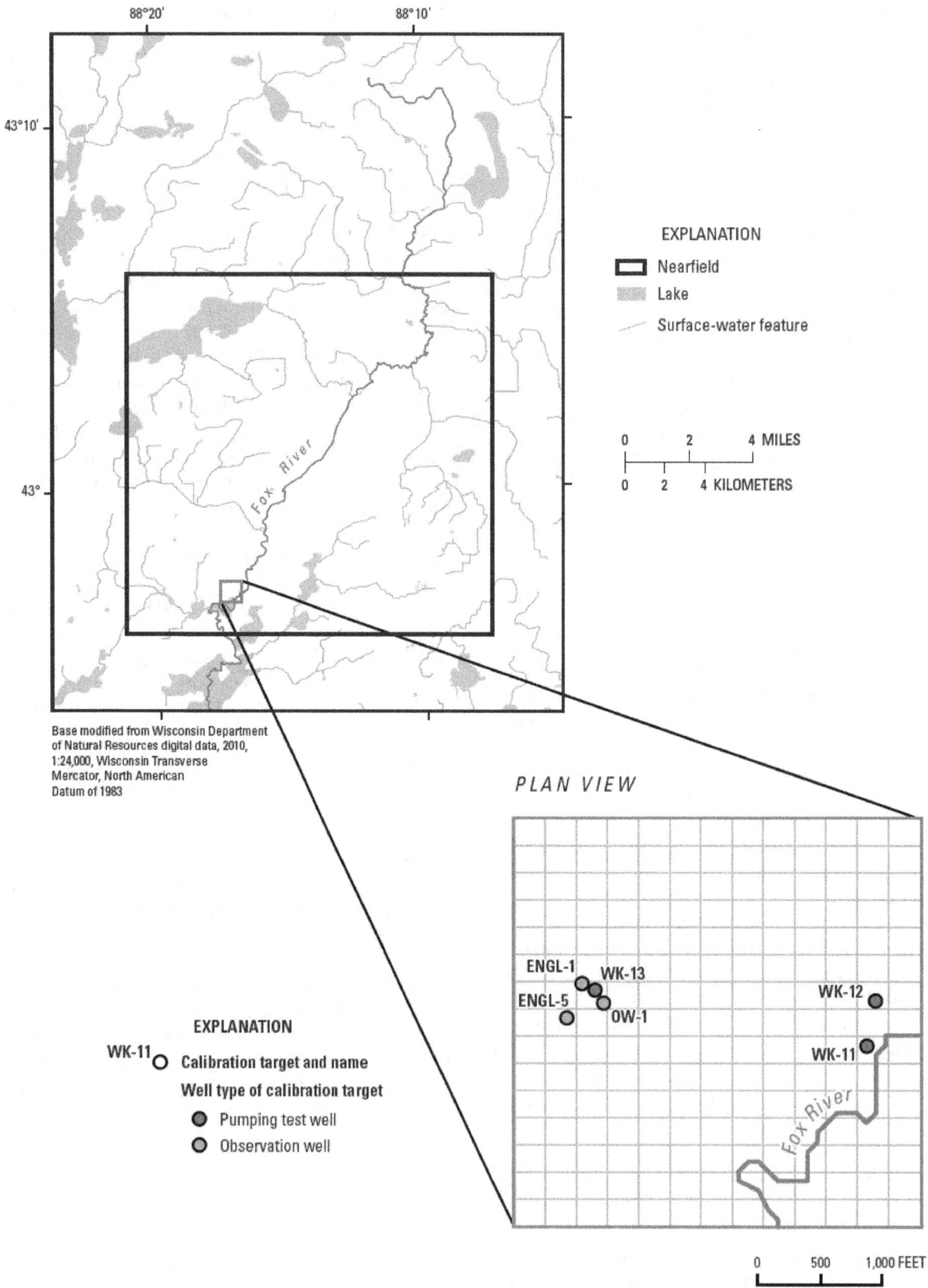

Figure 28*A*. Calibration targets: aquifer test location—plan view.

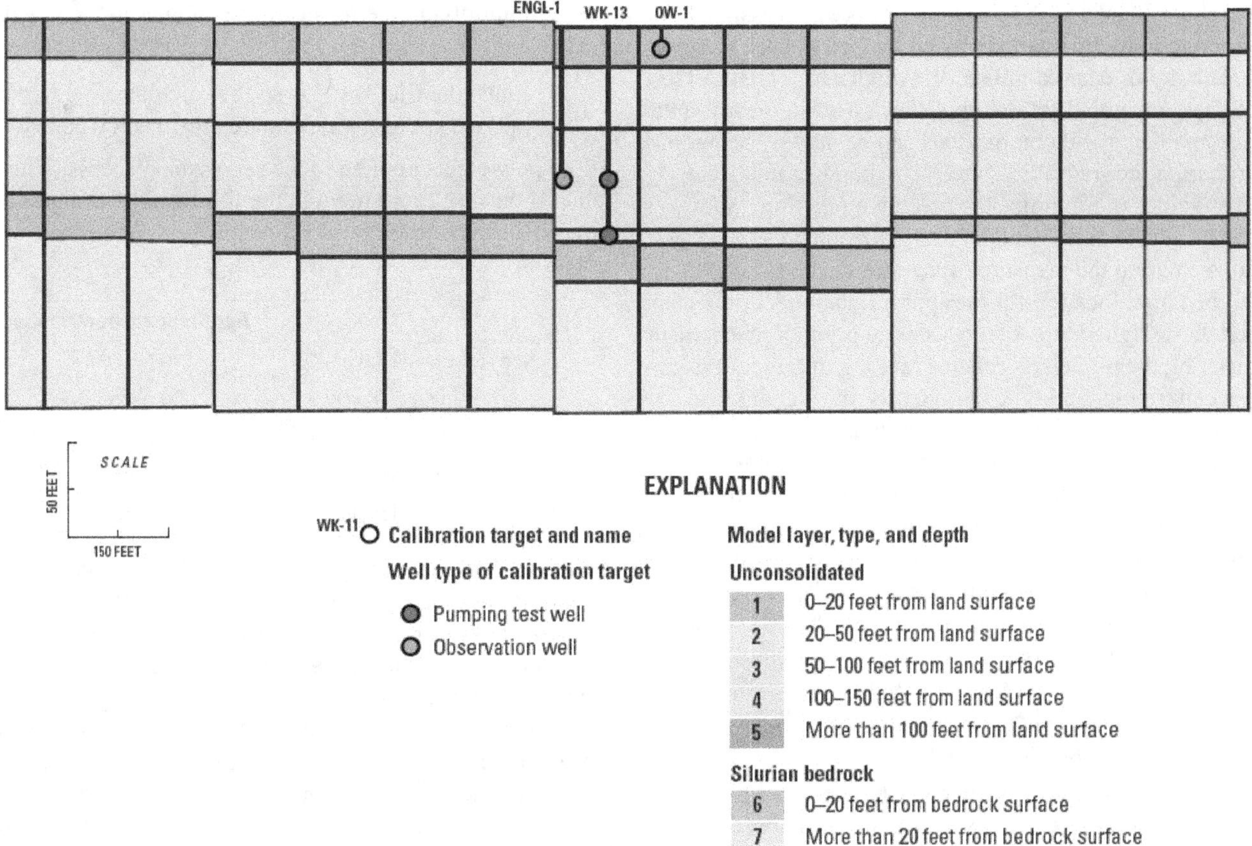

Figure 28*B*. Calibration targets: aquifer test location—vertical section.

Clearly, the model layering is not perfectly aligned with the local stratigraphic and well-construction conditions at WK-13, which limits the ability of the models to reproduce the drawdown trends. Moreover, it is unlikely that the mapping of only five facies to represent the flow properties of the unconsolidated aquifer over the entire Upper Fox River Basin is detailed enough to capture local features around a single well. Despite these obstacles, the aquifer test was added to the calibration process as a check on model construction in terms of the transmissivity implied by the response to the pumping at the WK-13 site. The transmissivity, which, along with vertical resistance and storage release, controls the drawdown response, is a function of the assemblage of facies near the pumping well at different depths for the two models. Also of interest was the models' ability to simulate the *absence* of response at OW-1, which serves as a constraint on the vertical hydraulic conductivity properties assigned the unconsolidated facies. Finally, simulation of the aquifer test in the calibration process allows estimation of storage properties of the shallow aquifer system at this one location within the basin. The simulated response is expected to be somewhat sensitive to the elastic aquifer properties represented by specific storage at the very beginning of the test and more sensitive to drainage properties represented by specific yield later in the test, but the ability to estimate specific yield is compromised by the delay in vertical leakage at the water table, which occurs when pumping is of short duration, and at screened intervals some distance below the water table. Starting values for the unconsolidated specific storage ($1x10^{-6}$ 1/ft) and specific yield (0.05) are based on preliminary analysis of the aquifer test data associated with pumping of WK-13 using curve-matching methods (Aquifer Science and Technology, 2008).

The model stress period devoted to the WK-13 aquifer test, lasting 20 hours, is divided into 12 time steps. The duration of the time steps increases geometrically by a factor of 1.4, such that the first step simulates the initial 0.1436 hours of the test and the last step simulates the final 5.917 hours of the test. For calibration, the drawdown at the times corresponding to the end of each time step is compared to the observed drawdown interpolated from the drawdown curve recorded during the test (based on approximately 250 observations at each observation well) The observed drawdown trends for the aquifer test at the two observation wells along with the match to simulated drawdown are shown in the "Model Calibration" section under "Calibration Results for Fine-Favored and Coarse-Favored models."

5.3 Target Weighting

Weights are used in the calibration process to assign importance to individual targets based on factors such as relative uncertainty or desired influence. Weights also are used to reconcile targets with different units (such as water levels and fluxes) so mathematical comparisons can be made. The parameter-estimation program PEST uses algorithms to modify parameter values to minimize the objective function; that is, the sum of squares of the weighted target residuals. The target residual is equal to the measured value minus the simulated value at the target location for the appropriate model time step. The modeler-assigned target weights are a primary mechanism for translating the modeler's relative ranking of target fit used to assess calibration quality; consequently, the weights are important for guiding the search for optimal parameters.

For this study, target weights were selected to normalize the influence of the target types by accounting for their uncertainty because of measurement uncertainty and structural uncertainty that impacts the reasonable expectations of how well the model is expected to reproduce them. This approach also serves to balance the contribution to the objective function among the target types to inform the estimation of different parameter sets (unconsolidated K_h and K_v zones, bedrock transmissivity and vertical resistance, streambed K, and storage terms). The method adopted was to

- choose the desired percent contribution of each target type to the objective function (sum of squares of weighted target residuals);
- select an expected residual for each target type (based on preliminary simulations with the initial parameters);
- calculate the contribution by multiplying the residual guess by the trial weight, squaring the product, and multiplying the result by the number in the target type; and
- adjust the trial weights so the calculated percent contributions approximately match the desired contributions.

The weights applied to the regression process (listed in table 9) were selected to yield the following ranking of target types in terms of their target contribution to the objective function:

	Approximate contribution
Nearfield water levels	40 percent
Vertical head differences	20 percent
Pumping test drawdown	15 percent
Flooding constraints	15 percent
Farfield water levels	10 percent

Nearfield water-level targets are collectively ranked most high in importance because they help estimate all the parameters (except storage terms) in the focus area of the study. Vertical head difference targets are ranked second highest because they guide the calibration process in optimizing the values of K_v parameters, otherwise difficult to estimate. The weights assigned the target types (for example, 1.0 in the case of nearfield water levels, 110 in the case of vertical head differences) reflect the relative importance and number of targets by set.

Table 9. Calibration target types and assignment of calibration weights.

[ft, foot; ft³/s, cubic foot per second; >, greater than; %, percent]

Target type	Group	Target value	Number of targets	Trial average absolute residual	Trial weight	Trial sum of squares of weighted residuals	Expected percent contribution to objective function
Nearfield water levels (ft)	1	variable	1,393	18	1	451,332	40.6
Farfield water levels (ft)	2	variable	2,099	18	.4	108,812	9.8
Base flow at Watertown and Waukesha gages (ft^3/s)	3	33.6 and 49.6	2	2	0	0	.0
Vertical head difference (ft)	4	variable	8	1.5	110	218,800	19.6
Pumping test drawdown at observation wells ENG-1 and ENGL-5 (ft)	5	variable	14	4	27	163,296	14.7
Pumping test drawdown at shallow well OW-1 (ft)	6	0.1	1	.5	0	0	.0
Minimize average flooding in flooded water table cells (ft)	7	0	1	7	42	86,436	7.8
Minimize percent nearfield cells without flooding > 3 ft (%)	8	0	1	9	32	82,944	7.5
Total						**1,110,620**	**100.0**

Two of the target types are assigned zero weight in the calibration process and make no contribution to the objective function: the base-flow estimates at the Watertown and Waukesha gages and the expected negligible drawdown at the end of the 20-hour aquifer test in shallow observation well OW-1. The base-flow targets primarily are affected by variations in recharge, and because the recharge zone values are kept fixed (see "Model Calibration" section under "Estimated Parameters"), it is convenient to simplify the inversion mathematics by assigning them zero weight. However, it is important to compare the simulated base flows to the estimated values after the optimization is complete as an independent check on the parameter estimation results. Similarly, the drawdown at OW-1 is neglected in the inversion to simplify the mathematical burden, but the quality of the calibration is evaluated with that target included. There is one other group of zero weights corresponding to drawdowns recorded at the two observation wells after 3.5 hours of pumping; the drawdown curves flatten considerably. This trend is due in large measure to the horizontal and vertical flow within the unconsolidated aquifer, but it is possible that it is also due in part to enhanced flow toward the well through fractures at the top of the Silurian dolomite. The pumping well screen is open to an elevation just above the top of the weathered dolomite, so that any conduits in the neighborhood could readily influence the test. Because of the uncertainty about these local conditions, it was judged more efficient to limit the active targets to the first 3.5 hours of the test; however, the comparison of simulated to observed drawdown is reported for the entire 20-hour test.

The fine-favored and coarse-favored models are calibrated using the same initial parameter values, the same solver input, and the same target types and weights. As discussed in previous sections, the calibration process yields distinct optimized values for the estimated parameters between the two models because they represent the geology differently.

5.4 Parameter Estimation Technique

Calibration of the Upper Fox River Basin models is tractable only if the parameter estimation process operates on an input structure that simplifies the complexity of the natural world. The required simplification operates at several levels. The initial model construction already radically simplifies real-world complexity by means of zones inside where values of hydraulic conductivity, recharge, streambed hydraulic conductivity, and storage are uniform. Zonation of inputs is a common simple form of "regularization" applied during model construction and calibration (Hunt and others, 2007a). However, there are other forms of regularization that can facilitate the inversion process. These other forms can be particularly important when, as in this study, the many parameters (even after zonation) are difficult to estimate. The use of the parameter-estimation computer code PEST (Doherty, 2008a; Doherty and Hunt and others, 2010) facilitates the application of sophisticated regularized inversion tools.

A groundwater-flow model with many parameters is commonly affected by parameter insensitivity and correlation, which in turn leads to solution non-uniqueness and an ill-posed inverse problem. In this study the "preferred value" form of Tikhonov regularization was used to counter these problems (Tikhonov 1963a, 1963b). The preferred values were set at the initial parameter values. Deviations from these values were penalized by increasing the sum of squares of weighted residuals, that is, the objective function. Therefore, deviations only occur if the regularization penalty is more than offset by a better match to targets and an overall smaller objective function. The amount of "preferred value" regularization applied to the fine-favored and coarse-favored models was relatively modest but served to exclude PEST results that minimized the objective function at the cost of unrealistic parameter values.

In contrast to Tikhonov regularization, which adds information to the calibration process in order to achieve numerical stability, subspace Singular Value Decomposition (SVD) regularization achieves numerical stability through subtracting parameters. This subtraction also is interpreted as the creation of a subset of parameters, each made up of a partial linear combination of all the parameters. As stated in Doherty and Hunt (2010, p. 4):

> "As a result of the subtraction, the calibration process is no longer required to estimate either individual parameters or combination of correlated parameters that are inestimable on the basis of the calibration dataset. These combinations are automatically determined through singular value decomposition (SVD) of the weighted Jacobian matrix. ...The Jacobian matrix consists of the sensitivities of all specified model outputs to all adjustable parameters. Individual parameters, or combinations of parameters, that are deemed to be estimable on the basis of the calibration dataset constitute the "calibration solution space." Those parameter/parameter combinations that are deemed to be inestimable (these constitute the "calibration null space") retain their initial values."

In this application, the Jacobian Matrix contains 17 columns corresponding to the estimated parameters listed in table 8, noting that the 18th entry, recharge, is fixed. Some of the parameters correspond not to a single zone but to multipliers on zones (for example, dolomite hydraulic conductivity). The rows in the Jacobian Matrix correspond to the individual targets summarized as types in table 9. Within PEST, the 17 estimated parameters (zone values or multipliers on initial values) were grouped into mathematical constructs representing combinations of parameters by means of the SVD algorithm. The fine-favored and coarse-favored models were each subjected to four PEST parameter-estimation updates after which the reduction in the objective functions became negligible or reversed. For both models, the number of estimated parameter combinations decreased from 17 to 12 in the course of the updates because of insensitivity of some combinations.

The fine-favored and coarse-favored updates were influenced not only by Tikhonov regularization and SVD subspace regularization, but also by lower and upper bounds placed on the values that parameters were allowed to assume (table 8). The K_h parameters are bound in such a way that during the PEST inversions, the fine-dominated facies would be assigned the smallest K_h value, the relatively fine facies the next smallest value (or a value equal to the fine-dominated facies), the mixed facies the next smallest value (or a value equal to the relatively fine facies), and so forth. This ordering ensures that the conceptual model underlying the computer model is enforced. A similar logic is used for the K_v parameters, except that only the fine-dominated facies K_v value is directly estimated. Because of the bigger spread in K_v values between zones, upper- and lower-bound ratios rather than values are used to enforce the ordering from lowest to highest K facies. A lower ratio bound equal to one allows the coarser facies in the ratio pair to be assigned the same K_v value as the finer facies, whereas, the upper ratio bound, always set to 100 for this parameter group, controls how much higher the coarse K_v can be relative to the finer K_v in the ratio pair. Upper and lower bounds determine acceptable values of the streambed hydraulic conductivity in such a way that, unlike the K_v ordering, allows overlap between the values of the three streambed zones (table 8). The bounds assigned the dolomite K and the storage terms are applied to the estimation of multipliers on initial values. Note that the exact same lower and upper bounds are applied to the inversions of the fine-favored and coarse-favored models.

The parameter-estimation problem was run on two 64-bit computers with multiple processors using the specialized parallel-processing version of PEST called BEOPEST (Schreüder, 2009; Hunt and others, 2010). Given the large number of times the models must be solved to implement the PEST algorithms, care was taken to reduce runtimes by (1) conducting preliminary runs with the fine-favored model to choose initial parameters that reproduce water-level targets reasonably well, (2) relaxing head and flux tolerances to the point where global mass balance errors for the runs equals about 0.1 percent, and (3) updating starting heads so that they correspond to preliminary simulations with initial parameter values. The total runtime for the fine-favored and coarse-favored inversions was 2 to 4 days.

5.5 Calibration Results for Fine-Favored and Coarse-Favored Models

The inversion routines in PEST yield improved congruence between observed and simulated values at targets by means of optimized parameter values. Comparison of observed to simulated targets shows improved fit with updates from initial and calibrated parameter values over most targets and target types for the fine-favored (table 10) and coarse-favored (table 11) models. The improved fit can be summarized by comparing the objective function contribution of each target type before and after optimization (table 12). The least target improvement occurs for the nearfield and farfield water levels for the fine-favored model—an expected result because preliminary manual calibration used to choose the initial parameter values was focused on the fine-favored model and was guided by the fit to water levels. Good target improvement for both models occurs for the vertical head difference and the aquifer test drawdown. The target types related to groundwater flooding show improvement for the fine-favored model (because of optimized parameter values that tend to lower the water table) but deterioration for the coarse-favored model (because of optimized parameter values that tend to raise the water table). Both models converge on thresholds that largely limit the flooding to riparian areas (see "Model Results" section).

The quality of fit for water levels is visualized by means of scatter plots showing the agreement over the range of head values (fig. 29) and by means of residual plots showing the spatial distribution of the errors (fig. 30). The vertical grouping of points (fig. 29) for the fine-grained and coarse-grained scatter plots is because of the use of water-level contours at 20-ft intervals as calibration targets. In figure 30, there is some spatial banding in the residuals for the fine-favored and coarse-favored models, which is linked to calibration error but may also be because of inaccuracies in the mapping of the water-level contours underlying the target values. Trial runs with the water-level targets moved from layers 2 and 4 to layer 1 reduced water-level residuals on an average of about 1 ft.

Table 10. Calibration fit for fine-favored model.

[Initial, simulation with initial parameter values; calibrated, results from optimal PEST iteration; residual, observed or estimated value less the simulated value; ft, foot; ft³/s, cubic foot per second; >, greater than; %, percent]

Water levels – statistics	Group	Mean error (ft)		Mean absolute error (ft)		Root mean square error (ft)	
		Initial	Calibrated	Initial	Calibrated	Initial	Calibrated
3,492 active targets	All	−5.17	−1.66	14.06	14.08	18.79	19.02
2,099 active targets	Farfield only	−6.77	−3.35	15.31	15.43	20.68	20.85
1,393 active targets	Nearfield only	−2.75	.89	12.12	12.05	15.52	15.89
Pumping test drawdown (ft)		**Observed**	**Initial**	**Calibrated**	**Residual**		
Observation well, ENG-1; Time steps 1–7							
active target	ENG-1-1	20.701	28.019	26.368	−5.667		
active target	ENG-1-2	26.795	31.409	31.572	−4.777		
active target	ENG-1-3	29.671	33.192	32.735	−3.064		
active target	ENG-1-4	31.847	34.549	33.528	−1.681		
active target	ENG-1-5	33.021	35.814	34.236	−1.215		
active target	ENG-1-6	34.396	37.14	34.993	−.597		
active target	ENG-1-7	35.096	38.639	35.851	−.755		
Observation well, ENG-1; Time steps 8–11							
inactive target	ENG-1-8	35.625	40.323	36.842	−1.217		
inactive target	ENG-1-9	36.035	42.503	37.923	−1.887		
inactive target	ENG-1-10	36.309	45.108	39.065	−2.757		
inactive target	ENG-1-11	36.39	48.18	40.138	−3.748		
Observation well, ENGL-5; Time steps 1–7							
active target	ENGL-5-1	13.236	7.169	12.529	0.707		
active target	ENGL-5-2	20.014	12.076	17.932	2.082		
active target	ENGL-5-3	23.354	15.192	20.148	3.206		
active target	ENGL-5-4	25.587	17.405	21.416	4.171		
active target	ENGL-5-5	26.847	19.149	22.274	4.573		
active target	ENGL-5-6	28.166	20.627	22.969	5.197		
active target	ENGL-5-7	28.895	21.968	23.607	5.287		

Table 10. Calibration fit for fine-favored model.—Continued

[Initial, simulation with initial parameter values; calibrated, results from optimal PEST iteration; residual, observed or estimated value less the simulated value; ft, foot; ft³/s, cubic foot per second; >, greater than; %, percent]

Pumping test drawdown (ft)		**Observed**	**Initial**	**Calibrated**	**Residual**
	Observation well, ENGL-5; Time steps 8–11				
inactive target	ENGL-5-8	29.485	23.093	24.28	5.205
inactive target	ENGL-5-9	29.897	24.511	24.971	4.926
inactive target	ENGL-5-10	30.184	26.011	25.722	4.462
inactive target	ENGL-5-11	30.384	27.697	26.416	3.968
	Shallow well				
		Assumed			
inactive target	OW-1	0.1	0.309	0.036	0.064
Base flow (ft³/s)		**Estimated**	**Initial**	**Calibrated**	**Fraction_target**
inactive target	Watertown Gage	33.6	32.1	32.0	0.95
inactive target	Waukesha Gage	49.6	45.1	45.4	.91
Vertical head difference (ft)		**Estimated**	**Initial**	**Calibrated**	**Residual**
active target	lath1	2.6	2.2	0.9	1.7
active target	lath3	−2.7	−1.2	−1.3	−1.4
active target	lath7	−2.4	−.8	−2.3	−.1
active target	lath8	−3.5	−1.1	−2.6	−.9
active target	lath9	−2.5	−.9	−2.3	−.2
active target	lath10	1	.8	−.4	1.4
active target	wk11	−4	−4.7	−4.8	.8
active target	wk12	−4	−5.1	−4.6	.6
Flooding		**Target**	**Initial**	**Calibrated**	
active target	Average flooding for flooded cells (ft)	minimize	7.7	6.5	
active target	Percent nearfield cells with flooding > 3 ft (%)	minimize	11.8	7.6	

Table 11. Calibration fit for coarse-favored model.

[Initial, simulation with initial parameter values; calibrated, results from optimal PEST iteration; residual, observed or estimated value less the simulated value; ft, foot; ft^3/d, cubic foot per day; >, greater than; %, percent]

Water levels – statistics	Group	Mean error (ft)		Mean absolute error (ft)		Root mean square error (ft)	
		Initial	Calibrated	Initial	Calibrated	Initial	Calibrated
3,492 active targets	All	1.76	−0.60	14.41	14.03	19.72	18.91
2,099 active targets	Farfield only	−1.40	−3.02	14.96	15.05	20.40	20.19
1,393 active targets	Nearfield only	6.52	3.04	13.60	12.48	18.65	16.82

Pumping test drawdown (ft)		Observed	Initial	Calibrated	Residual
	Observation well, ENG-1; Time steps 1–7				
active target	ENG-1-1	20.701	27.382	24.382	−3.681
active target	ENG-1-2	26.795	30.676	30.424	−3.628
active target	ENG-1-3	29.671	32.35	33.151	−3.479
active target	ENG-1-4	31.847	33.537	34.487	−2.64
active target	ENG-1-5	33.021	34.563	35.122	−2.101
active target	ENG-1-6	34.396	35.611	35.282	−.885
active target	ENG-1-7	35.096	36.823	35.314	−.219
	Observation well, ENG-1; Time steps 8–11				
inactive target	ENG-1-8	35.625	38.263	35.332	0.293
inactive target	ENG-1-9	36.035	40.144	35.35	.686
inactive target	ENG-1-10	36.309	42.418	35.372	.937
inactive target	ENG-1-11	36.39	45.096	35.401	.988
	Observation well, ENGL-5; Time steps 1–7				
active target	ENGL-5-1	13.236	7	10.777	2.459
active target	ENGL-5-2	20.014	11.757	16.212	3.803
active target	ENGL-5-3	23.354	14.699	19.026	4.328
active target	ENGL-5-4	25.587	16.658	20.481	5.106
active target	ENGL-5-5	26.847	18.038	21.198	5.649
active target	ENGL-5-6	28.166	19.083	21.453	6.713
active target	ENGL-5-7	28.895	19.982	21.542	7.353

Table 11. Calibration fit for coarse-favored model.—Continued

[Initial, simulation with initial parameter values; calibrated, results from optimal PEST iteration; residual, observed or estimated value less the simulated value; ft, foot; ft³/d, cubic foot per day; >, greater than; %, percent]

Pumping test drawdown (ft)		Observed	Initial	Calibrated	Residual
		Observation well, ENGL-5; Time steps 8–11			
inactive target	ENGL-5-8	29.485	20.771	21.581	7.904
inactive target	ENGL-5-9	29.897	21.819	21.61	8.287
inactive target	ENGL-5-10	30.184	22.985	21.642	8.542
inactive target	ENGL-5-11	30.384	24.328	21.68	8.704
		Shallow well			
inactive target	OW-1	Assumed 0.1	0.067	0.105	−0.005
Base flow (ft³/d)		**Observed**	**Initial**	**Calibrated**	**Fraction_target**
inactive target	Watertown Gage	33.6	31.9	32.5	0.97
inactive target	Waukesha Gage	49.6	46.5	46.5	.94
Vertical head difference (ft)		**Observed**	**Initial**	**Calibrated**	**Residual**
active target	lath1	2.6	1.4	0.9	1.7
active target	lath3	−2.7	.3	−1.2	−1.5
active target	lath7	−2.4	.1	−2.6	.2
active target	lath8	−3.5	−.3	−3.0	−.5
active target	lath9	−2.5	.2	−2.6	.1
active target	lath10	1	1.3	−.3	1.3
active target	wk11	−4	−2.8	−4.0	.0
active target	wk12	−4	−3.4	−4.5	.5
Flooding		**Target**	**Initial**	**Calibrated**	
active target	Average flooding for flooded cells (ft)	minimize	5.8	6.7	
active target	Percent nearfield cells with flooding > 3 ft (%)	minimize	4.1	6.1	

Table 12. Objective function contributions for initial and final calibrated models.

[Initial values correspond to contributions before PEST inversion; final values correspond to contributions for calibrated model after successive PEST inversions; contributions to objective function are dimensionless; NA, not applicable; >, greater than; ft, foot]

Target type	Group	Initial contribution to weighted objective function	Final contribution to weighted objective function	Percent change in objective function contribution
Fine-favored model				
Nearfield water levels	1	335,431	351,750	5 (increase)
Farfield water levels	2	143,638	145,930	2 increase
Base flow at Watertown and Waukesha gages	3	0	0	NA
Vertical head difference	4	185,387	104,200	−44 (reduction)
Pumping test drawdown at observation wells ENG-1 and ENGL-5	5	378,990	129,710	−66 (reduction)
Pumping test drawdown at shallow well OW-1	6	0	0	NA
Minimize average flooding in flooded water table cells	7	105,093	75,076	−29 (reduction)
Minimize percent nearfield cells with flooding > 3 ft	8	142,119	58,693	−59 (reduction)
TOTAL		**1,290,660**	**865,380**	**−33 (reduction)**

Target type	Group	Initial contribution to weighted objective function	Final contribution to weighted objective function	Percent change in objective function contribution
Coarse-favored model				
Nearfield water levels	1	484,524	394,070	−19 (reduction)
Farfield water levels	2	139,802	136,840	−2 reduction
Base flow at Watertown and Waukesha gages	3	0	0	NA
Vertical head difference	4	430,446	91,321	−79 (reduction)
Pumping test drawdown at observation wells ENG-1 and ENGL-5	5	421,234	180,340	−57 (reduction)
Pumping test drawdown at shallow well OW-1	6	0	0	NA
Minimize average flooding in flooded water table cells	7	59,611	78,248	31 (increase)
Minimize percent nearfield cells with flooding > 3 ft	8	17,052	37,828	122 (increase)
TOTAL		**1,552,670**	**918,650**	**−41 (reduction)**

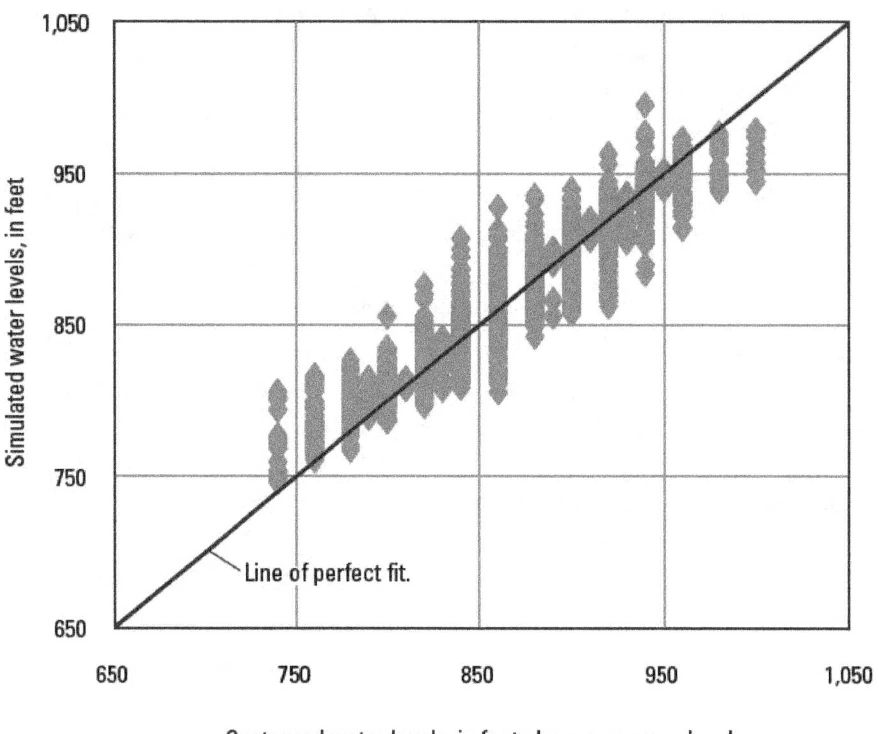

Figure 29*A*. Calibration scatter plots for water levels—fine-favored model.

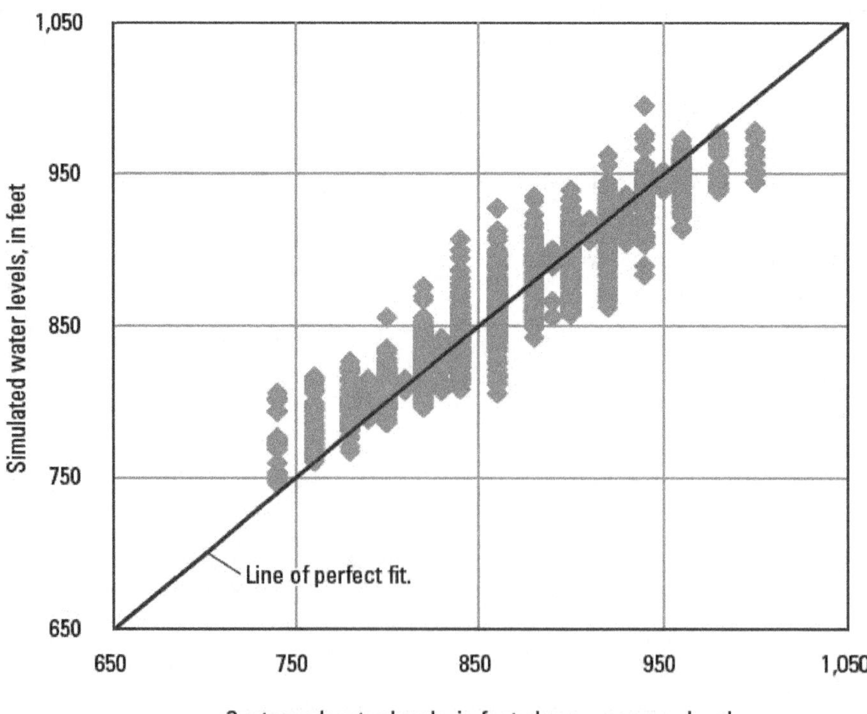

Figure 29*B*. Calibration scatter plots for water levels—coarse-favored model.

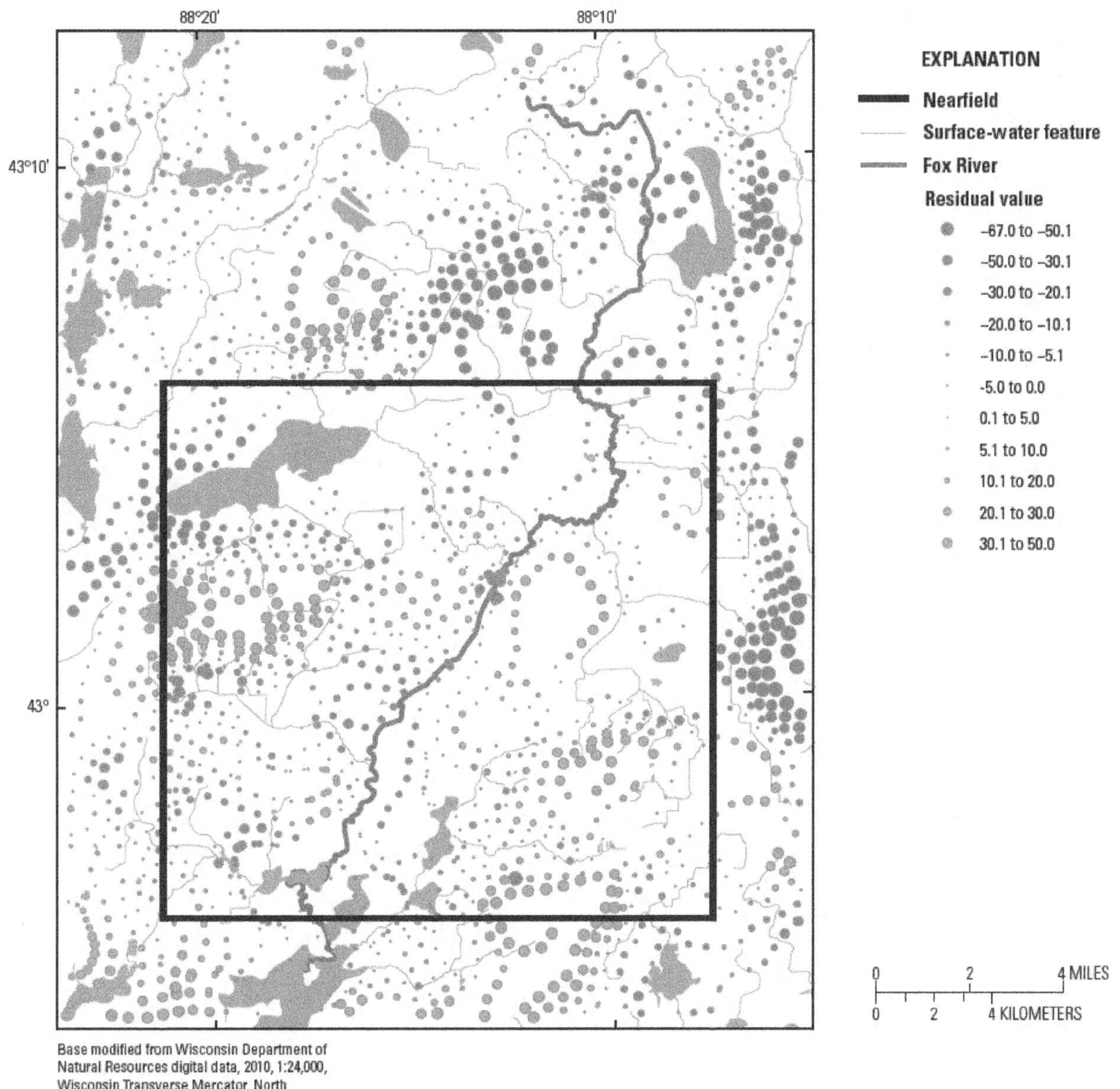

Figure 30A. Spatial distribution of water-level residuals (feet)—fine-favored model. (Residual = observed − simulated; blue residuals indicate observed greater than simulated; red residuals indicate observed less than simulated.)

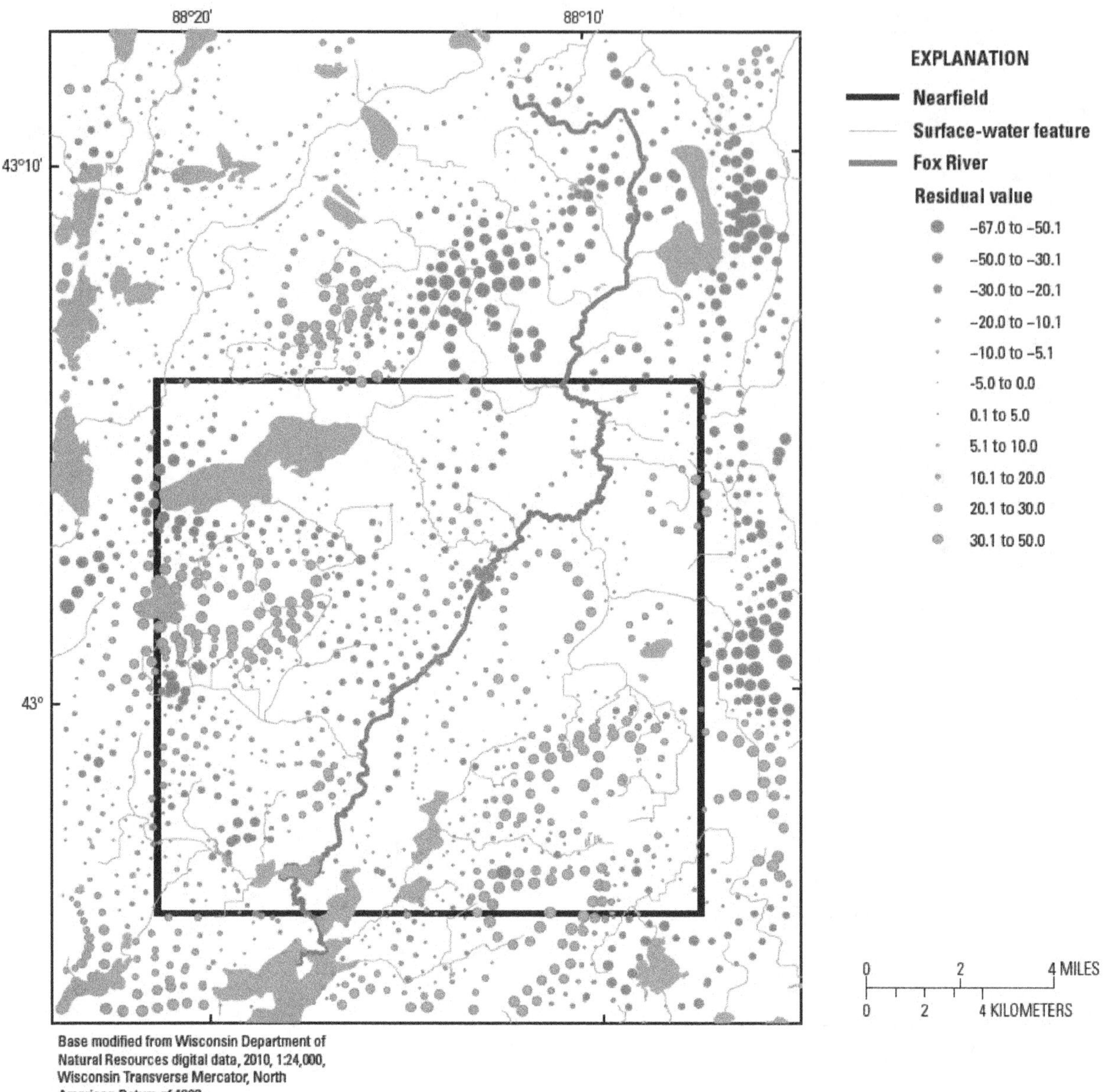

Figure 30*B*. Spatial distribution of water-level residuals (feet)—coarse-favored model. (Residual = observed − simulated; blue residuals indicate observed greater than simulated; red residuals indicate observed less than simulated.)

The calibration scatter plots and the residual distribution figures both indicate that low water levels associated with valleys tend to be overestimated in the simulations and high water levels associated with uplands tend to be underestimated. Despite the calibration process this bias indicates lingering deficiencies in the model. However, it is also possible that part of the bias is due to the target values themselves, which might reflect perched conditions under uplands. It is also possible that the water-level contour map (from which the targets are derived) systematically underestimates the water table elevation in valleys (where data density is low) because assumed contours neglect the tendency for groundwater discharge and high water-table elevations to occur over riparian areas larger than the width of the stream channels.

The congruence between the observed and simulated drawdown at aquifer test observation wells (fig. 31) shows fairly good agreement for both models in the early part of the test (first 3.5 hours). It is speculated that local heterogeneity in the dolomite could be affecting the drawdown slope in the latter part of the test. In any event, the comparison indicates that the overall transmissivity and vertical resistance applied to both models in the vicinity of pumping well WK-13 is not greatly in error even though the model was not constructed to precisely match the hydrogeology in this area. Another check on model performance is the simulated drawdown at shallow well OW-1, located in layer 1 of the same row and column as the well pumping 600 gal/min from layers 3 and 4 . For the optimized parameter values, the simulated drawdown at OW-1 after 20 hrs of pumping is 0.04 and 0.1 ft in the fine-favored and coarse-favored models, respectively (tables 10 and 11), whereas, the simulated drawdown adjacent to the pumping well opening, located less than 100 ft away, is more than 60 ft. Spot measurements during the test showed no drawdown at OW-1 (as opposed to drawdown on the order of 30 ft at the observation wells), which indicates that the stress from the well does not propagate in the vertical direction nearly as easily as it does in the lateral direction. The model appears to be in agreement with this finding.

The match for the vertical head difference targets (tables 10 and 11) can be summarized in terms of how well observed "flowing" conditions at six of the eight target test wells (open on average 100 ft below land surface but reporting heads above land surface) are reproduced by the models. Measured water levels in the six flowing wells were 1 to 3 ft above land surface. In the fine-favored and coarse-favored model, five of these six wells are simulated as flowing with water levels 0.5 to 2.5 ft above land surface.

The second dimension of the calibration results are the optimized parameter values generated by the inversion routines (table 13). Partly because of the imposed bounds, the results honor the conceptual model in terms of the ordering of K_h and K_v values by facies zones. The coarse-favored K values are generally lower than the fine-favored K values. This is reasonable because both models are calibrated to the same targets, but the coarse-favored model contains higher K material than the fine-favored model and, in compensation, requires lower K values to obtain similar results, whereas, the calibration of the fine-favored model, given its relatively larger volume of fine-grained material, tends to yield relatively higher K values to meet target values. For both models,facies 1, 2, and 3 have appreciably lower K_v values than facies 4, which, in turn, is markedly lower than facies 5. This result is welcome because it restricts vertical preferential flow to areas where facies 4 and, especially, facies 5 are present. The implied vertical anisotropy results are a useful check on the calibration process. For the initial values, vertical anisotropy was set lowest for the dominantly fine and dominantly coarse facies and set highest for the mixed fine and coarse facies. This ordering follows the logic that the contrast between K_h and K_v should be weakest for facies dominated by a single texture and strongest for the facies with the most varied textures. The calibrated K values generally imply the same ordering for the fine-favored and coarse-favored models. The most striking departure from the initial settings is the large size of the implied vertical anisotropy for the mixed facies (table 13).

The optimized results for streambed K are higher in most cases than the initial values across the two models but still respect the ordering from low to middle to high K zones. Calibration of both models yielded lower transmissivity and higher vertical conductivity for the dolomite bedrock by factors averaging about 1.5 (table 13). The optimized specific storage results from the WK-13 aquifer test are very close to the global specific storage assigned the subsurface in the regional SEWRPC model, $2.6e^{-7}$ 1/ft (Feinstein and others, 2005a). The optimized specific yield values varied greatly between the fine-favored and coarse-favored models (table 13), which suggests that limiting the analysis to the first 3.5 hrs of the aquifer test was insufficient time to generate reliable estimates of specific yield by means of the optimizations.

One test of the quality of the calibration process is to examine the number of parameters where calibrated values equal lower or upper bounds. The less constrained the optimization is by bounds, the more it is informed by mathematical criteria of best fit, and the less it is informed by prior information (that is judgment) imposed by the user. Among the 17 parameters estimated, the fine-favored calibration returns three upper bounds and two lower bounds, whereas, the coarse-favored calibration returns five lower bounds (all involving K) and one upper bound (table 13). These results suggest that without the imposition of bounds, the objective function could have been lower in both numerical models but at the expense of violating some elements of the underlying conceptual model.

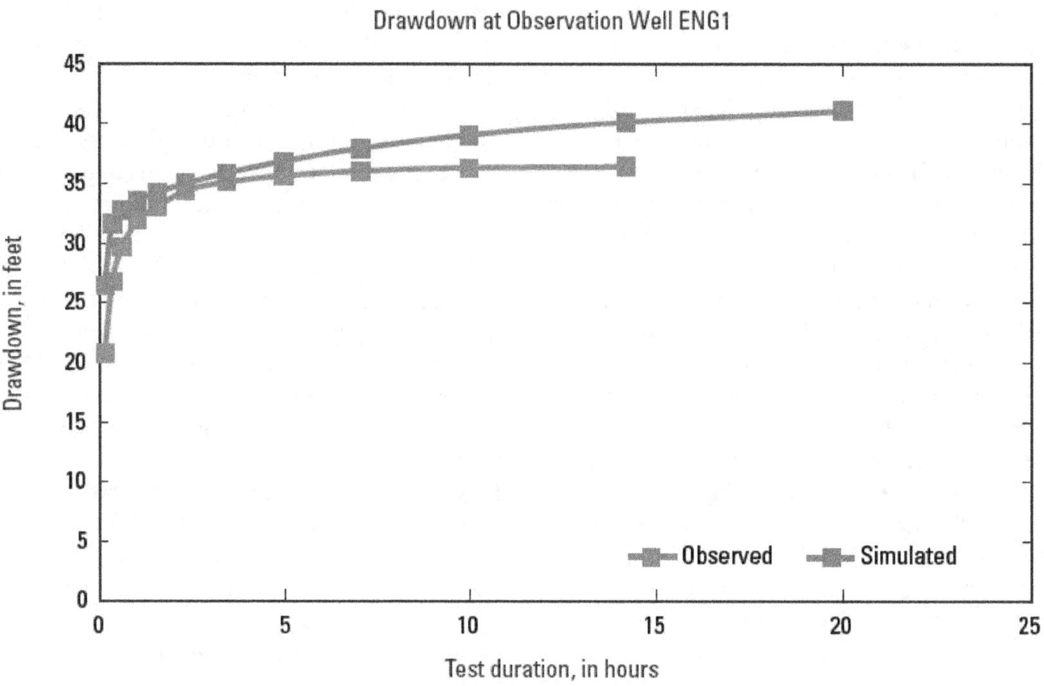

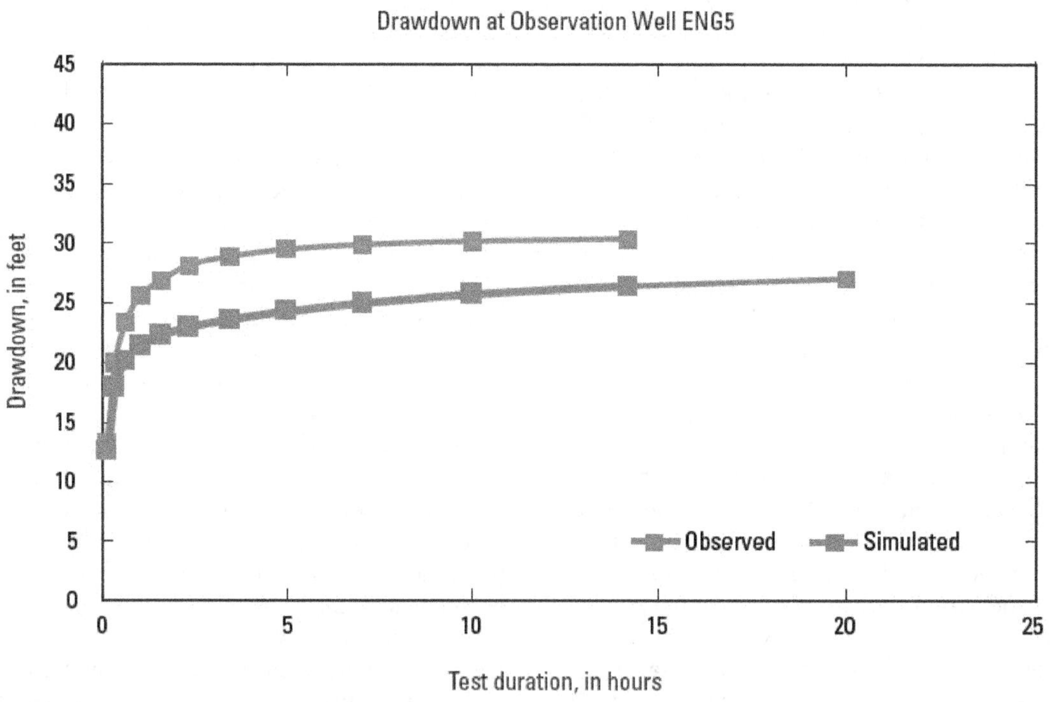

Figure 31*A*. Observed and simulated drawdown at pumping-test observation wells—fine-favored model.

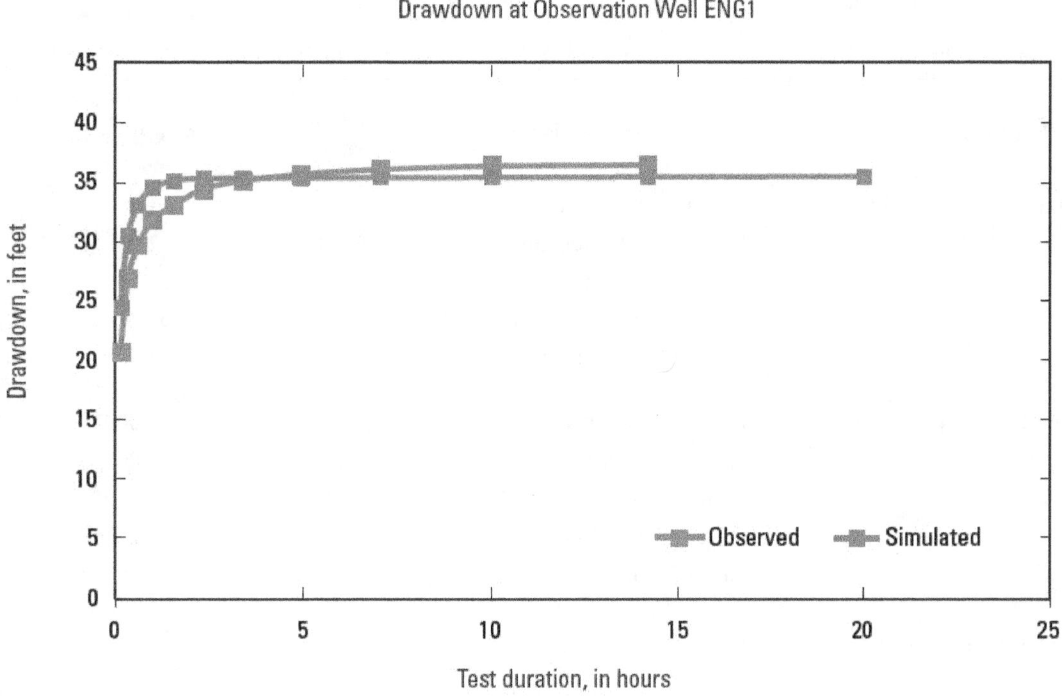

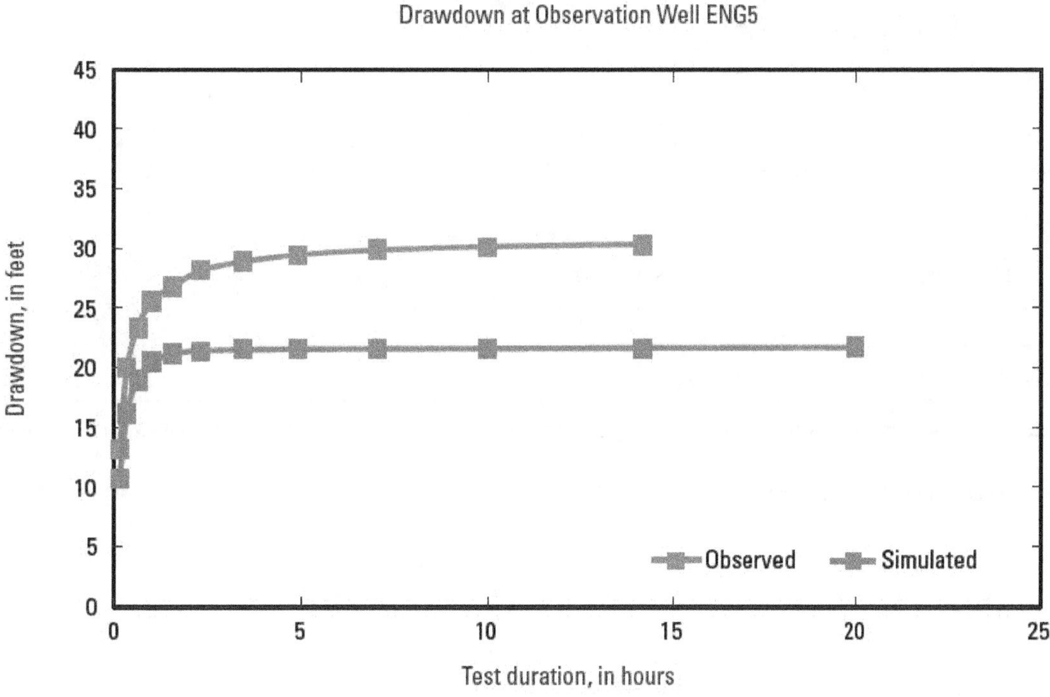

Figure 31*B*. Observed and simulated drawdown at pumping-test observation wells—coarse-favored model.

Table 13. Calibrated parameter values for fine- and coarse-favored models.

[ft, foot; ft/d, foot per day; kh, horizontal hydraulic conductivity; kv, vertical hydraulic conductivity; ksfr, streambed hydraulic conductivity]

Parameter set	Zone or variable	Initial value	Calibrated value	
			Fine-favored model	Coarse-favored model
Horizontal hydraulic conductivity (ft/d)	Facies 1 – dominantly fine	0.5	[2]1	0.75
	Facies 2 – relatively fine	2	2.26	[2]1
	Facies 3 – mixed fine and coarse	10	[2]20	19.75
	Facies 4 – relatively coarse	40	51.31	[2]20
	Facies 5 – dominantly coarse	80	183.09	[2]60
Vertical hydraulic conductivity (ft/d)[1]	Facies 1 – dominantly fine	0.005	0.00955	[2]0.002
	Facies 2 – relatively fine	.01	[2].00955	.00245
	Facies 3 – mixed fine and coarse	.02	[2].00955	[2].00245
	Facies 4 – relatively coarse	.2	.24361	.12658
	Facies 5 – dominantly coarse	2	1.66107	7.08675
Implied vertical anisotropy (k_h:k_v)	Facies 1 – dominantly fine	100	105	376
	Facies 2 – relatively fine	200	237	408
	Facies 3 – mixed fine and coarse	500	2,095	8,062
	Facies 4 – relatively coarse	200	211	158
	Facies 5 – dominantly coarse	40	110	8
Streambed hydraulic conductivity (ft/d)	ksfr_low	1	0.8	2.0
	ksfr_middle	5	9.6	[2]25
	ksfr_high	25	[2]100	63.6
Multiplier on dolomite hydraulic conductivity	kh	1	0.77	0.42
	kv	1	1.48	1.55
Storage of unconsolidated deposits	specific storage (ft^{-1})	1.00E-06	2.86E-07	3.68E-07
	specific yield (--)	.05	.028	.306
Multiplier on recharge (fixed)	recharge	1	1	1

[1]For vertical hydraulic conductivity, facies 1 was estimated directly and the remaining parameters were estimated using sequential multipliers bounded between 1 and 100.

[2]Calibrated value at bound set in PEST inversion.

Another common check on the quality of the calibration process involves determining that the amount of correlation between parameters is not large enough to subvert the ability to independently estimate their values. However, the use of SVD subspace methods in the inversion routines invalidates calculation of correlation coefficients, so that check cannot be made.

Finally, it is noteworthy that the calibrated K_h values are systematically higher for the fine-favored model than for the coarse-favored model. In fact, if all the nearfield hydraulic conductivity values weighted by the saturated thickness in glacial cells are summed and then divided by the sum of the saturated thicknesses, the resulting global K_h value across the five glacial layers for the fine-favored model is about 20-percent higher than the global value for the coarse-favored model (26 and 22 ft/d, respectively). The difference in the average values does not mean that groundwater is transmitted more easily to surface water and wells in the fine-favored than the coarse-favored model given the crucial role that the facies distribution plays in controlling flow. Because the coarse model imposes more continuity among the coarse facies, it supports more preferential flow in the horizontal and vertical directions even though the K_h values assigned to those facies are moderately lower than the values assigned in the fine-favored model. Results from simulations of a hypothetical pumping from wells along the Fox River are consistent with this conclusion (see "Model Application to Hypothetical Well Field" section).

5.6 Sensitivity Analyses

The PEST code generates composite sensitivity values between all observations and individual estimated parameters for each update of the parameter values. These sensitivity values indicate how well parameters can be estimated from the available information. Composite sensitivities compiled at the beginning of the inversion and for the final update (fig. 32 for fine-favored model and fig. 33 for coarse-favored model) indicate that the targets inform K_h, K_v, dolomite transmissivity and vertical resistance, and specific storage better than streambed K or specific yield. The parameter most sensitive to observations is the K_h of the mixed facies. The relative composite sensitivity by parameter set for the final inversion update is as follows:

Parameter set	Fine-favored model (percent)	Coarse-favored model (percent)
K_h	39	51
K_v	46	36
Streambed K	3	2
Storage terms	6	6
Dolomite K	6	5

This tabulation indicates that for the final update K_v was the most sensitive target type for the fine-favored calibration, but K_h was the most sensitive for the coarse-favored calibration. The analysis also shows that streambed K values are insensitive to the assembled targets. It is interesting to note that initial PEST runs, which included the base-flow targets along with recharge as a parameter, produced even less sensitivity with respect to streambed conductance than did the final PEST implementation without any base-flow targets or recharge parameters. Given the uncertainty attached to the streambed K values, further analysis is needed to explore how changing the values can influence model results (see "Model Application to Hypothetical Well Field" section).

PEST output also can be used to look at the sensitivity relation to determine the most informative target types by parameter set. The rankings are similar for the two models and show that all the active target types participated in the optimizations:

Parameter set	Most informative target types
K_h	most sets, especially water levels and aquifer-test drawdowns
K_v	vertical head differences
Streambed K	flooding constraints and nearfield water levels
Storage terms	aquifer-test drawdowns
Dolomite K	water levels

EXPLANATION

Abbreviation	Description
kx2	horizontal hydraulic conductivity of unlithified facies 1 (fine dominated)
kx4	horizontal hydraulic conductivity of unlithified facies 2 (relatively fine)
kx12	horizontal hydraulic conductivity of unlithified facies 3 (mixed)
kx13	horizontal hydraulic conductivity of unlithified facies 4 (relatively coarse)
kx14	horizontal hydraulic conductivity of unlithified facies 5 (coarse dominated)
kz2	vertical hydraulic conductivity of unlithified facies 1
rat4_2	vertical hydraulic conductivity of unlithified facies 2 expressed as ratio to facies 1
rat12_4	vertical hydraulic conductivity of unlithified facies 3 expressed as ratio to facies 2
kx13	vertical hydraulic conductivity of unlithified facies 4 expressed as ratio to facies 3
kx14	vertical hydraulic conductivity of unlithified facies 5 expressed as ratio to facies 4
ksfrl	hydraulic conductivity for streambeds characterized as muddy, silty
ksfrm	hydraulic conductivity for uncharacterized streambeds
ksfrh	hydraulic conductivity for streambeds characterized as sandy and gravelly
ssm	specific storage of unlithified deposits (model layers 1–5) expressed as single multiplier
sym	specific yield of unlithified deposits (model layers 1–5) expressed as single multiplier
kxsimul	horizontal hydraulic conductivity of Silurian dolomite expressed as single multiplier on 8 zones
kzsimul	vertical hydraulic conductivity of Silurian dolomite expressed as single multiplier on 8 zones

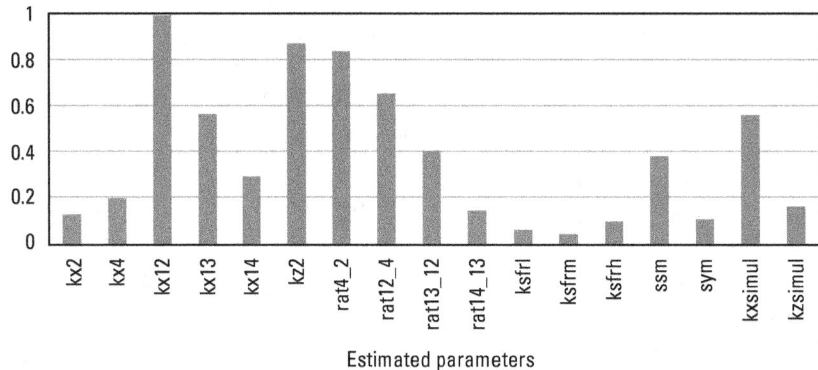

Figure 32*A*. Composite sensitivities to calibration targets for fine-favored model generated by PEST—initial parameter values.

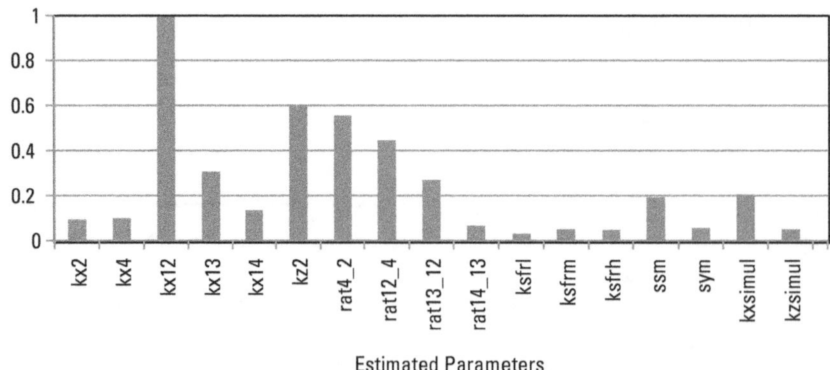

Figure 32*B*. Composite sensitivities to calibration targets for fine-favored model generated by PEST—calibrated parameter values.

EXPLANATION

Abbreviation	Description
kx2	horizontal hydraulic conductivity of unlithified facies 1 (fine dominated)
kx4	horizontal hydraulic conductivity of unlithified facies 2 (relatively fine)
kx12	horizontal hydraulic conductivity of unlithified facies 3 (mixed)
kx13	horizontal hydraulic conductivity of unlithified facies 4 (relatively coarse)
kx14	horizontal hydraulic conductivity of unlithified facies 5 (coarse dominated)
kz2	vertical hydraulic conductivity of unlithified facies 1
rat4_2	vertical hydraulic conductivity of unlithified facies 2 expressed as ratio to facies 1
rat12_4	vertical hydraulic conductivity of unlithified facies 3 expressed as ratio to facies 2
kx13	vertical hydraulic conductivity of unlithified facies 4 expressed as ratio to facies 3
kx14	vertical hydraulic conductivity of unlithified facies 5 expressed as ratio to facies 4
ksfrl	hydraulic conductivity for streambeds characterized as muddy, silty
ksfrm	hydraulic conductivity for uncharacterized streambeds
ksfrh	hydraulic conductivity for streambeds characterized as sandy and gravelly
ssm	specific storage of unlithified deposits (model layers 1–5) expressed as single multiplier
sym	specific yield of unlithified deposits (model layers 1–5) expressed as single multiplier
kxsimul	horizontal hydraulic conductivity of Silurian dolomite expressed as single multiplier on 8 zones
kzsimul	vertical hydraulic conductivity of Silurian dolomite expressed as single multiplier on 8 zones

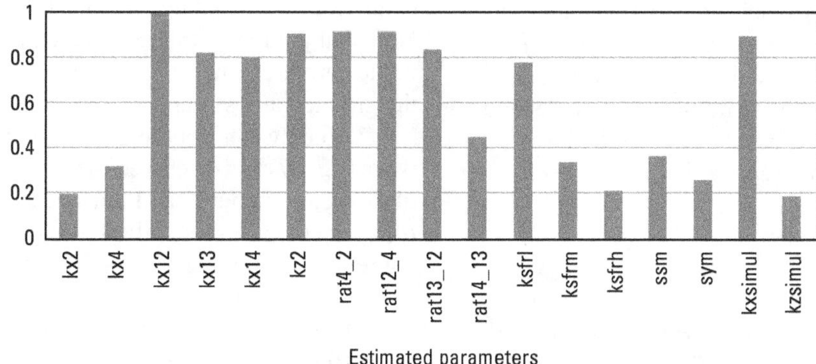

Figure 33*A*. Composite sensitivities to calibration targets for coarse-favored model generated by PEST—initial parameter values.

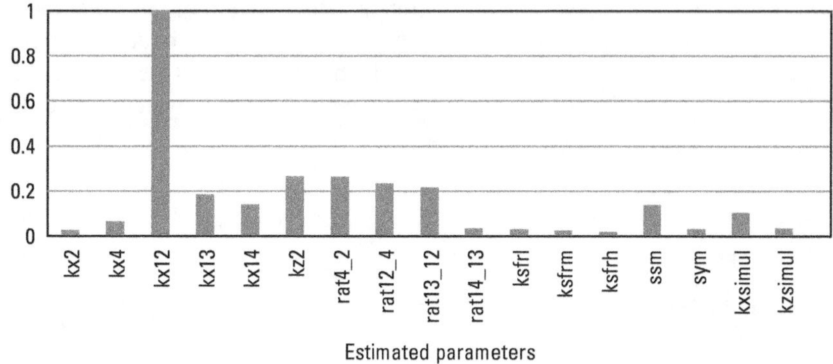

Figure 33*B*. Composite sensitivities to calibration targets for coarse-favored model generated by PEST—calibrated parameter values.

6. Model Results

The calibrated versions of the fine-favored and coarse-favored models simulate water levels including the water-table elevation, water budgets, and contributing areas for discharge zones such as streams, pumping wells, and quarries. These results are based on steady-state simulations intended to approximate 2005 conditions. In both models, the numerical solutions achieve favorable balance between sources and sinks. The global mass balance error for the steady-state stress period is 0.05 percent for the fine-favored model (convergence after 173 outer iterations of the solver) and 0.04 percent for the coarse-favored model (convergence after 771 outer iterations of the solver). Checks of the mass balance errors in subdomains of the models suggest that local errors are of the same order as the global error.

6.1 Water Levels for Fine-Favored and Coarse-Favored Models

As pointed out in the "Model Solver" section of this report, it is the merit of the MODFLOW-NWT formulation that dry cells remain active, so that abrupt changes in the vertical flow profile leading to numerically isolated cells and spikes in water-table elevation are avoided (fig. 24*B*). Accordingly, the water-table solutions are acceptably smooth (fig. 34). The average depth from land surface to the water table across the model domain is 39.12 ft for the fine-favored model and 38.7 ft for the coarse-favored model (within model layer 2). The percent distribution of the water table by layer depth interval is similar for the two models:

Layer depth interval	Fine-favored model (percent)	Coarse-favored model (percent)
1 (0 to 20-ft depth)	50.6	51.5
2 (20.1- to 5-ft depth)	21.7	22.2
3 (50.1- to 100-ft depth)	19.9	20.0
4 (100.1- to 150-ft depth)	6.5	5.6
5 (more than 150.1-ft depth)	.7	.4
6 (upper 20 ft of bedrock)	.6	.3

The percentage of water-table cells simulated as perched (locations where a water-table cell is underlain by a dry cell) is very small: 0.04 percent for the fine-favored model and 0.02 percent for the coarse-favored model. All the water levels simulated in perched cells are reasonable elevations (unlike the unacceptably high elevations simulated for many more perched cells using PCG2).

Areas of vertically upward gradients in the unconsolidated deposits are generally simulated along surface-water bodies (fig. 35). The deeper parts of the unconsolidated material are pressurized to a fairly high degree in places and upward head differences over 10 ft are common. This result sorts well with the observation that water levels in test wells open to the bottom of the unconsolidated material are often flowing (John Jansen, formerly Aquifer & Technology, oral commun., February 2010) when it is assumed that the location of flowing wells is typically in low-lying areas and that the water table is close to the land surface. Heterogeneity in the hydraulic conductivity field combines with the geometry of discharge zones and the simulated head distribution to produce a complicated pattern of upward flow to the water table (fig. 36). Most areas of upward flow are simulated to transmit groundwater at a rate between 0.0001 and 0.001 ft/d, but the upward flow rate transmits 0.01 ft/d or more in some riparian reaches. These results demonstrate the ability of the models to simulate preferential flow associated with natural discharge. Note that pumping stresses inserted in the models can also induce preferential flow away from surface-water features into the groundwater system.

Groundwater flooding is simulated mostly along waterways (fig. 37). As discussed in the "Model Calibration" section of this report under "Land Surface Constraints", this outcome is expected because not all existing riparian discharge zones are represented in the models, causing the groundwater in the models to discharge directly to SFR and LAK cells rather than to adjacent riparian zones. By contrast, in upland areas of relatively low water-table levels, discharge to the stream channel is limited and, as a result, base flow does not accumulate along the channel. In this case, the stream can be simulated as dry for low-flow conditions (ephemeral). For headwater reaches this condition commonly occurs when the water-table elevation is below the top of the streambed. Ephemeral streams are simulated to occur in about 20 percent of the stream reaches inside the Upper Fox River Basin, representing approximately 40 percent of the headwater reaches (fig. 37).

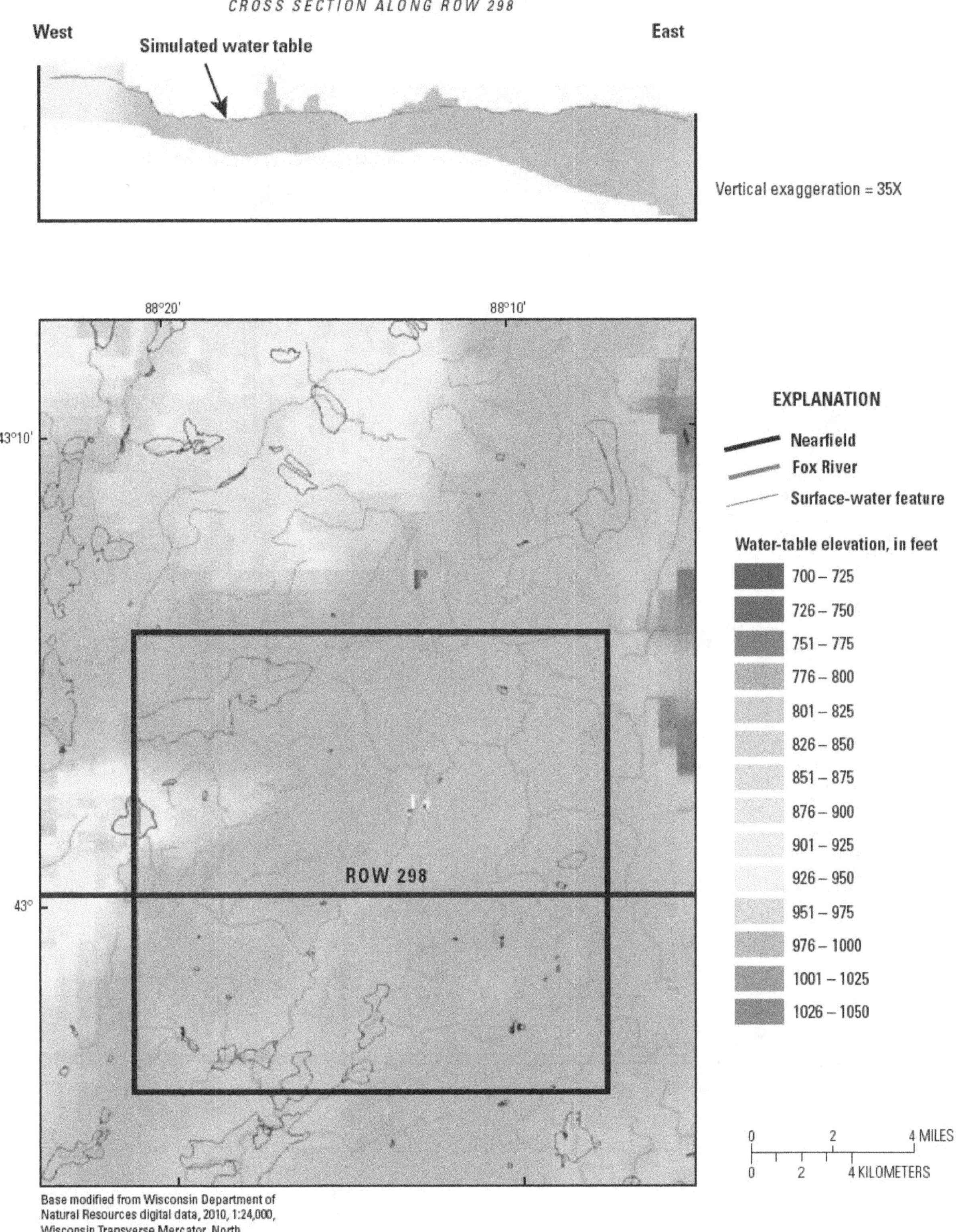

Figure 34*A*. Simulated water table in plan view and vertical section—fine-favored model. Vertical cross section along row 298 of the model. Vertical exaggeration is 35 times.

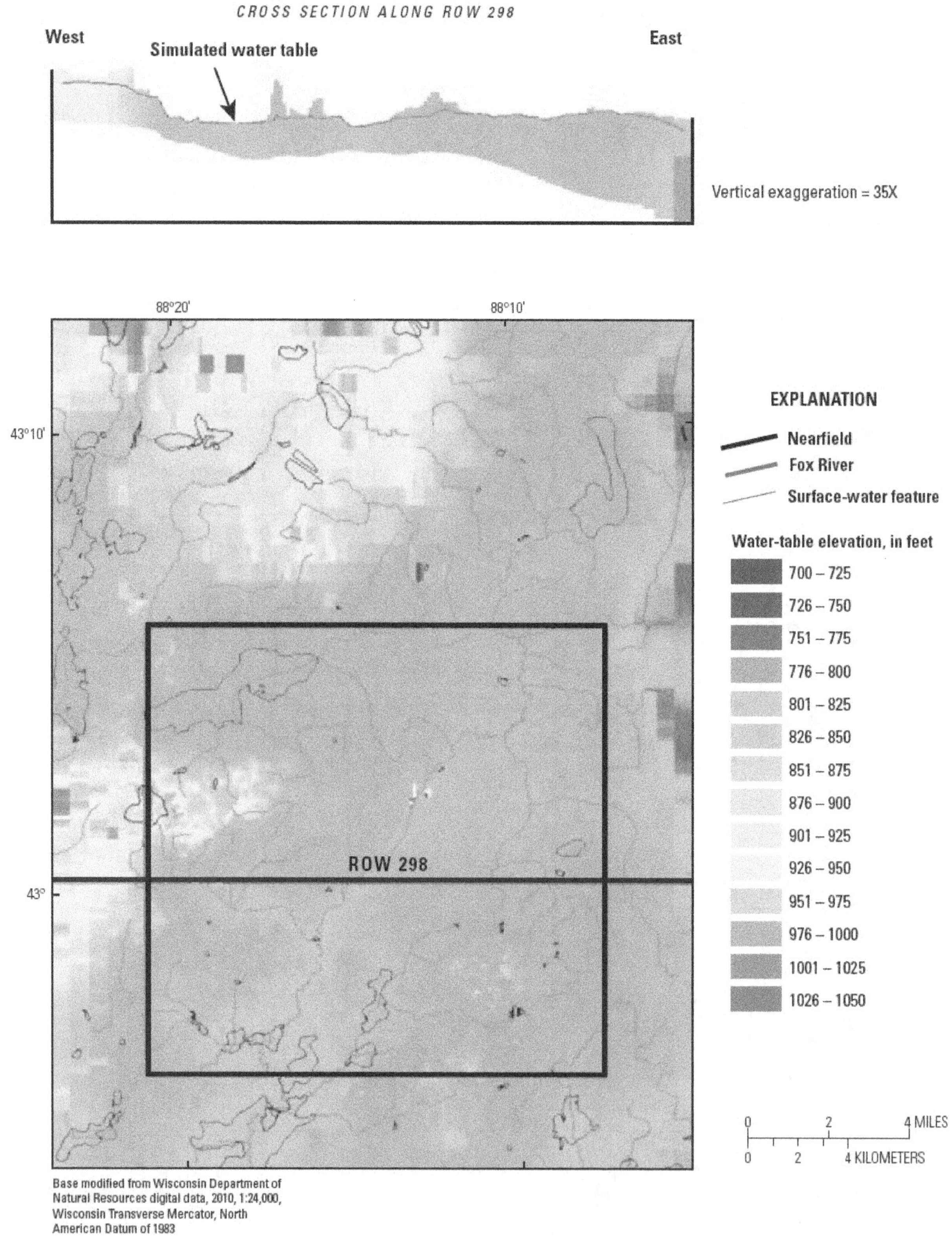

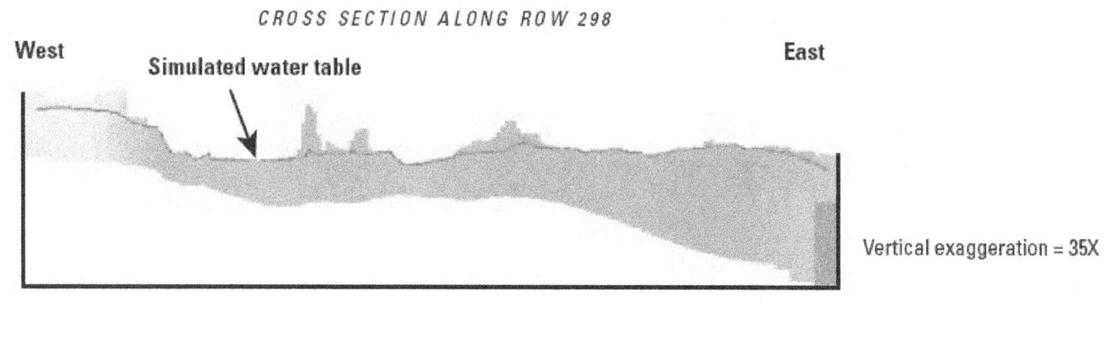

88°20'
88°10'
43°10'
43°
ROW 298

EXPLANATION

Nearfield
Fox River
Surface-water feature

Water-table elevation, in feet

700 – 725
726 – 750
751 – 775
776 – 800
801 – 825
826 – 850
851 – 875
876 – 900
901 – 925
926 – 950
951 – 975
976 – 1000
1001 – 1025
1026 – 1050

0 2 4 MILES
0 2 4 KILOMETERS

Base modified from Wisconsin Department of Natural Resources digital data, 2010, 1:24,000, Wisconsin Transverse Mercator, North American Datum of 1983

Figure 34*B*. Simulated water table in plan view and vertical section—coarse-favored model. Vertical cross section along row 298 of the model. Vertical exaggeration is 35 times.

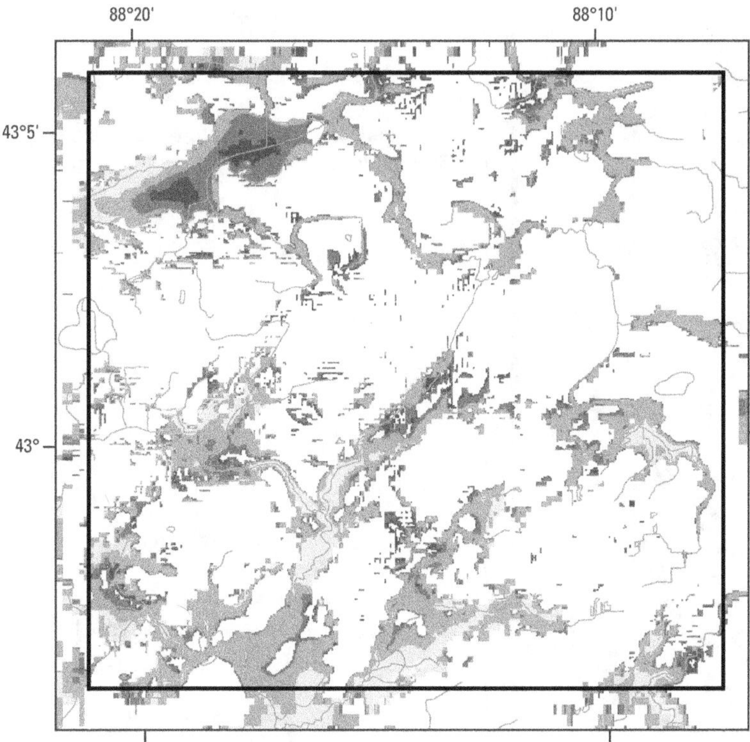

Figure 35*A*. Areas of simulated upward head gradients in unconsolidated deposits for model nearfield—fine-favored model. White area is where head gradient is downward in unconsolidated deposits.

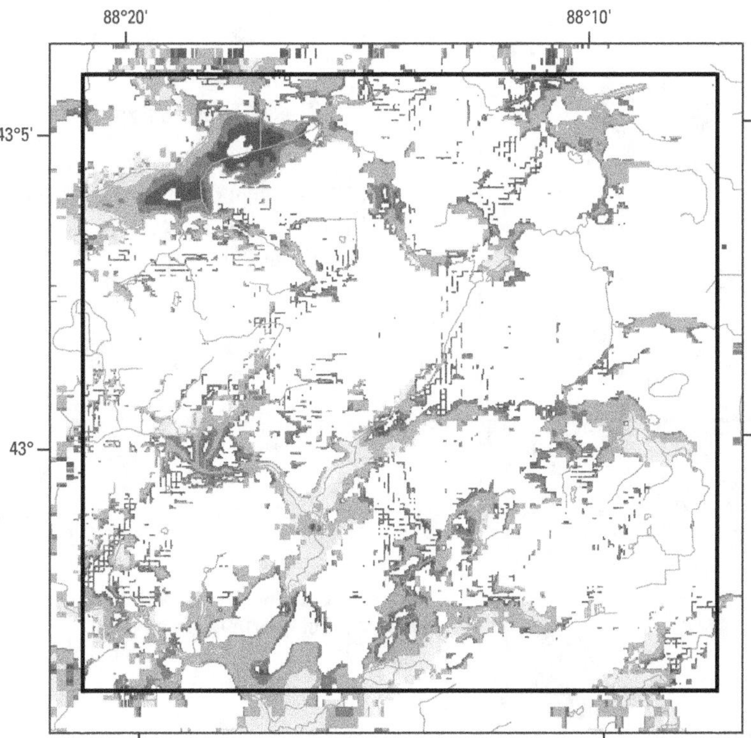

Figure 35*B*. Areas of simulated upward head gradients in unconsolidated deposits for model nearfield—coarse-favored model. White area is where head gradient is downward in unconsolidated deposits.

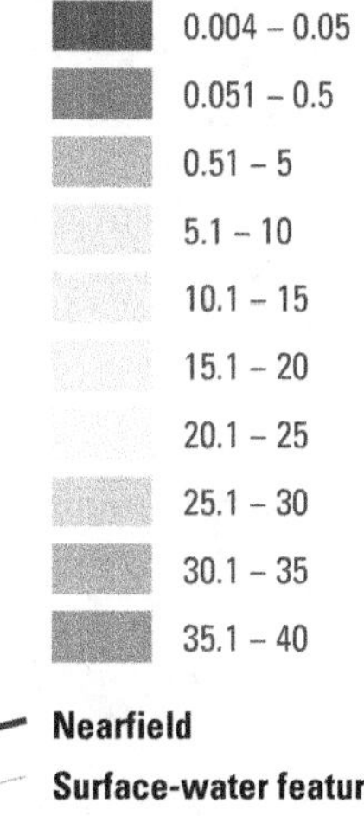

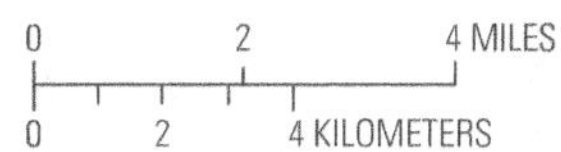

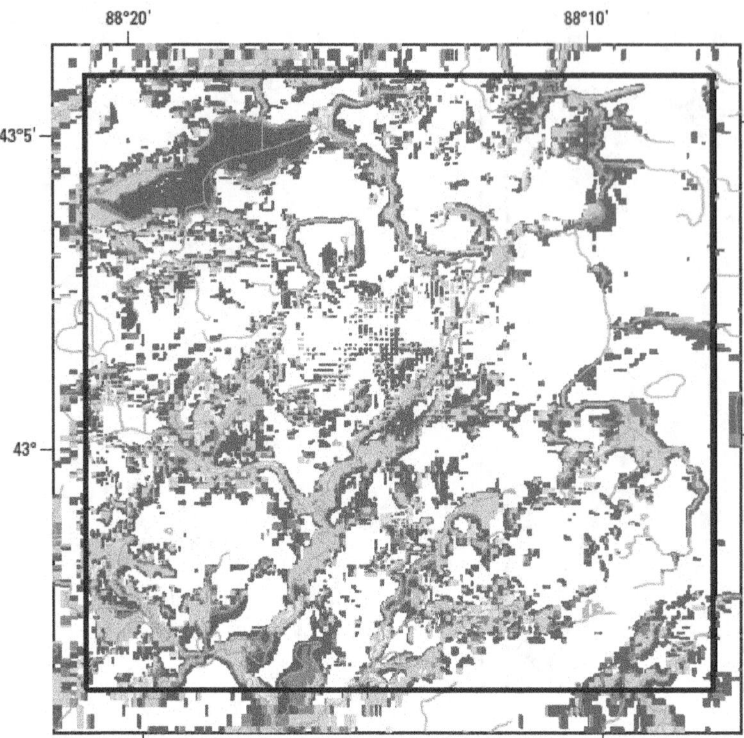

a) Areas of upward flow to water table = 34.6 percent of nearfield

Figure 36*A*. Areas of upward flow to water table for model nearfield—fine-favored model. White area is where flow is downward from water table.

EXPLANATION

Upward flow rate, in feet per day, to unperched water-table layer

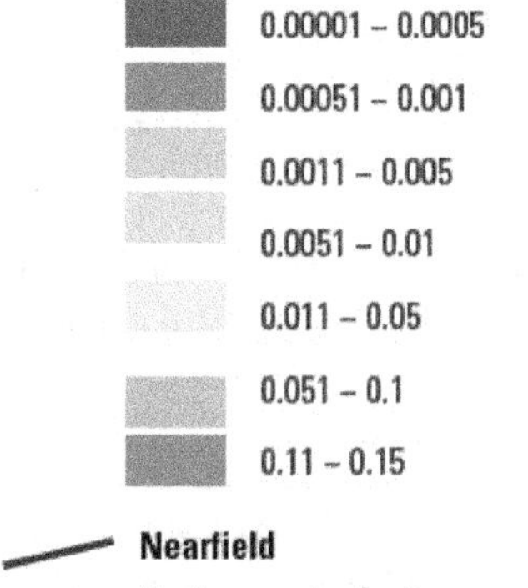

Upward flow to water table corresponds to flow to unperched water-table cell (whatever layer water table is in) from underlying cell

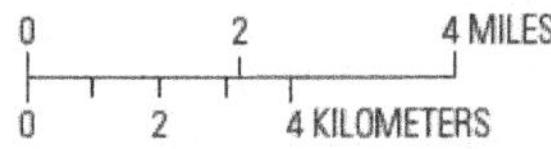

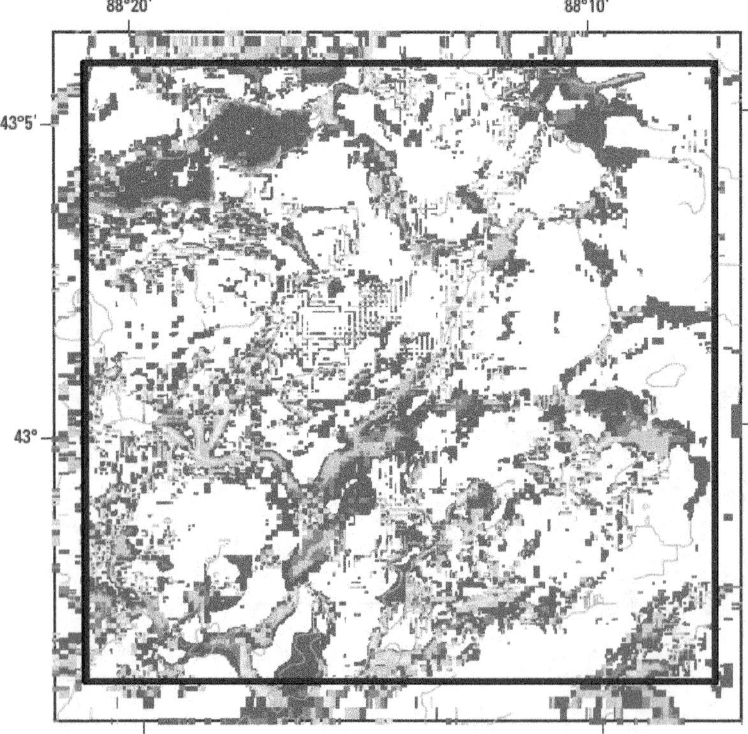

b) Areas of upward flow to water table = 33.8 percent of nearfield

Figure 36*B*. Areas of upward flow to water table for model nearfield—coarse-favored model. White area is where flow is downward from water table.

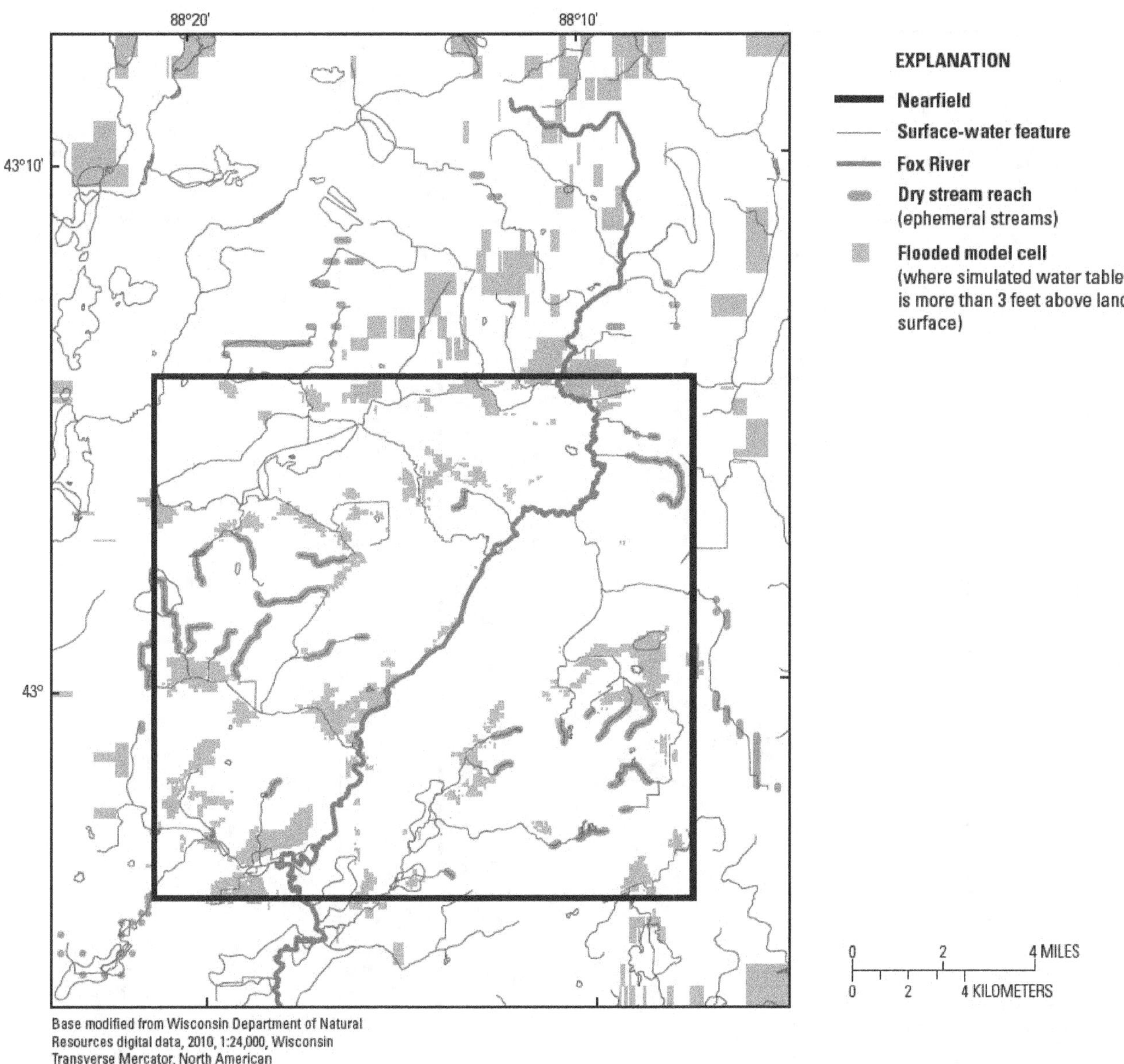

Figure 37A. Simulated flooded water-table cells and dry headwater reaches for model domain—fine-favored model. [18.8 percent (1,457 out of 7,731) of stream reaches are dry; all are headwater reaches.]

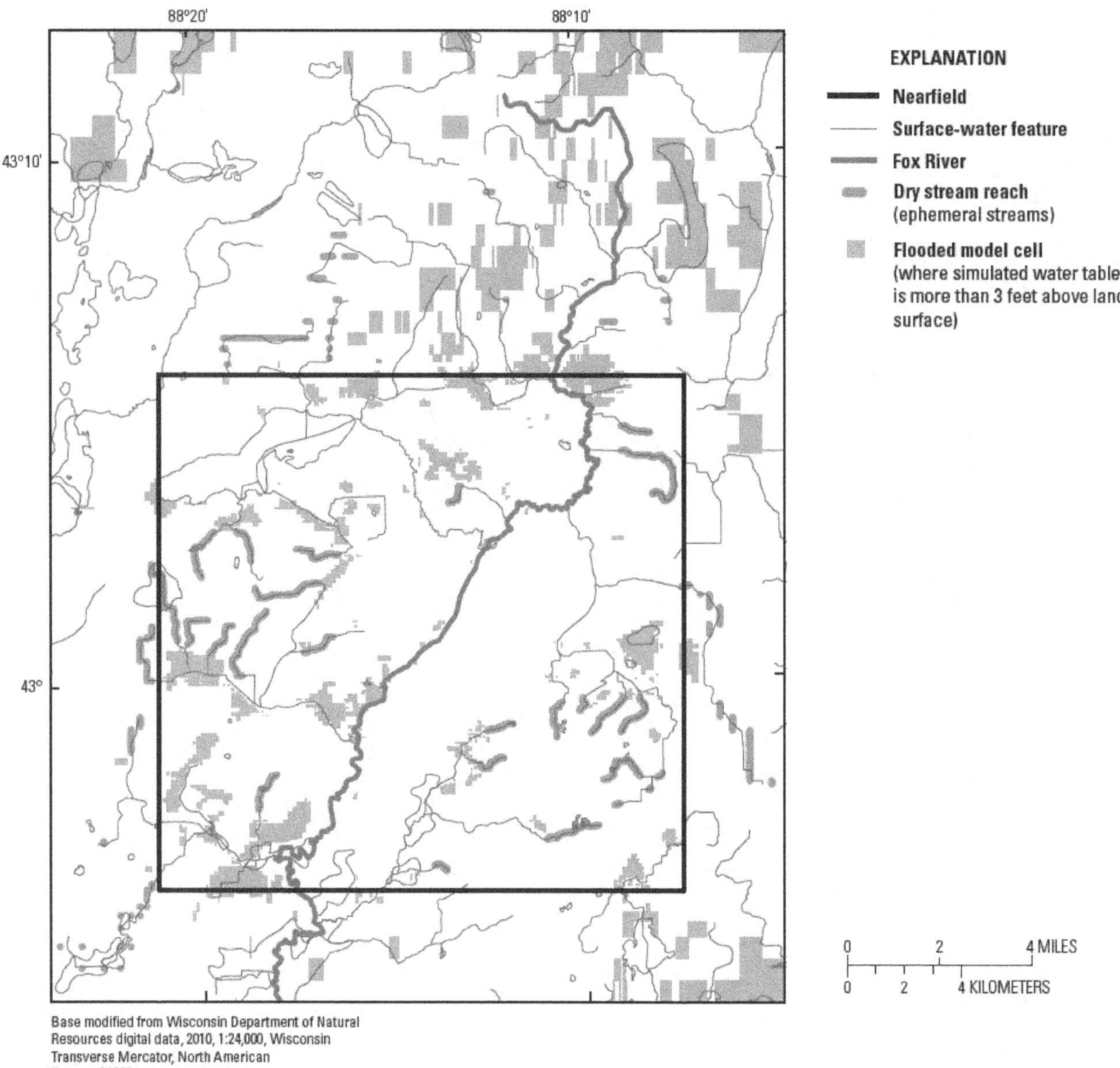

Figure 37*B*. Simulated flooded water-table cells and dry headwater reaches for model domain—coarse-favored model. [21.2 percent (1,638 out of 7,731) of stream reaches are dry; all but 10 are headwater reaches.]

6.2 Water Budgets for Fine-Favored and Coarse-Favored Models

The water budgets simulated by the calibrated Upper Fox models combine prescribed inputs (such as boundary fluxes and recharge) with calculated outputs (such as base flow and quarry discharge). The steady-state fluxes associated with the sources (recharge, inflow across model boundaries, and inflow to the groundwater system from surface-water features) balance the steady-state fluxes associated with sinks (outflow across model boundaries, well discharge, quarry discharge, and base flow to surface-water features). The relative magnitudes of the simulated inflows and outflows are very similar for the fine-favored and coarse-favored models. For the model domain as a whole, recharge is the dominant inflow and base flow is the dominant outflow with each term representing over 80 percent of the sources and sinks (table 14). The amount of groundwater discharge to streams within the Upper Fox River Basin is about nine times greater than the amount of water induced from streams to the groundwater system for the fine-favored model and about seven times greater for the coarse-favored model. Pumping from the high-capacity wells included in the models is responsible for most of the induced flow and totals about 7 percent of the simulated domain outflow.

Table 14. Simulated water budgets for model domain.

[Mgal/d, million gallons per day; for flows related to surface water, MODFLOW package simulating flow is indicated in parentheses]

	Fine-favored model		Coarse-favored model	
	Flux (Mgal/d)	**Percent of total**	**Flux (Mgal/d)**	**Percent of total**
Inflows = sources				
Recharge	72.3	82.9	72.3	82.9
Boundary fluxes	1.6	1.8	1.6	1.9
Lateral (model grid edges)	1.3	1.5	1.3	1.5
Bottom (Silurian/Maquoketa contact)	.3	.3	.3	.5
Induced flow from surface water	13.4	15.3	13.3	15.2
Upper Fox River Basin streams (SFR)	3.9	4.5	5.1	5.8
Other streams (RIV)	9.4	10.7	8.1	9.3
Pewaukee Lake (LAK)	.1	.1	.1	.1
Other lakes and wetlands (DRN)	.0	.0	.0	.0
TOTAL	87.3	100.0	87.3	100.0
Outflows = sinks				
Boundary fluxes	8.9	10.2	8.9	10.2
Lateral (model grid edges)	5.4	6.2	5.4	6.2
Bottom (Silurian/Maquoketa contact)	3.5	4.0	3.5	4.0
High Capacity Wells	6.4	7.3	6.1	7.0
Unconsolidated (13 wells)	1.6	1.8	1.6	1.8
Dolomite (68 wells)	4.8	5.5	4.5	5.2
Sussex and Waukesha Quarries	1.3	1.5	1.0	1.2
Base flow to surface water	70.8	81.0	71.2	81.6
Upper Fox River Basin streams (SFR)	34.7	39.8	37.3	42.8
Other streams (RIV)	23.0	26.2	23.3	26.6
Pewaukee Lake (LAK)	1.6	1.9	1.7	1.9
Other lakes and wetlands (DRN)	11.5	13.2	9.0	10.3
TOTAL	87.4	100.0	87.2	100.0

For this application, the NWT solver decreases pumping when the saturated thickness of a cell hosting a well falls below 20 percent of its total thickness. From a total of 6.7 Mgal/d input to the models, the loss of pumping because of this condition is less than 4 percent for the fine-favored model and less than 8 percent for the coarse-favored model. Almost all the loss is from Silurian dolomite wells. The discrepancy between what withdrawals the calibrated parameters can support and the pumping rates the public-supply systems have reported might point to calibration errors, but it could also be a function of overestimated pumping rates. It is interesting to note that preliminary simulations using MODFLOW-2005 with the PCG2 solver yielded losses of pumping discharge because of dry (inactive) cells equal to almost twice the amount lost in the MODFLOW-NWT simulations as a result of reduced saturated thickness.

Water budgets calculated for nested basins show a trend in the relative magnitude of outflows to sinks with size of basin. The areas corresponding to the basins upgradient from the Watertown and the Waukesha streamgages (fig. 6*B*) define nested subbasins within the Upper Fox River Basin. For the fine-favored (table 15) and coarse-favored (table 16) models, the proportion of outflow that circulates to surface-water features increases with basin size, accompanied by a decrease in the proportion of well discharge. This outcome is partly linked to the concentration of pumping wells northeast of Waukesha inside the Watertown subbasin nested inside the other two basins. The simulated budget for the entire Upper Fox River Basin for the two models indicates that about 2 percent of the groundwater flow leaks downward from the basin into the deep part of the flow system, 12 to 15 percent moves laterally across the basin boundaries, 77 to 80 percent discharges to Upper Fox surface waters, and about 7 percent to pumping wells within the basin.

Table 15. Simulated water budgets for nested basins in the fine-favored model, Upper Fox River Basin.

[Mgal/d, million gallons per day; ΣIN, sum of inflows; mi^2, square mile; see figure 7 for outlines of basins]

	Inflow (Mgal/d)	Outflow (Mgal/d)	OUT percent of ΣIN
Upper Fox River Basin upgradient of Watertown gage (area is 86 mi^2)			
Recharge	16.63		
Vertical flux across Silurian/Maquoketa boundary	.00	0.27	1.4
Lateral flux across basin boundaries	1.83	5.12	25.7
Surface-water exchange[1]	1.43	11.58	58.2
Pumping		2.99	15.0
			100.3
			[mass balance error = 0.3]
Upper Fox River Basin upgradient of Waukesha gage (area is 135 mi^2)			
Recharge	24.34		
Vertical flux across Silurian/Maquoketa boundary	.10	0.57	1.9
Lateral flux across basin boundaries	4.09	6.28	20.6
Surface-water exchange[1]	2.00	20.19	66.1
Pumping		3.55	11.6
			100.2
			[mass balance error = 0.2]
Upper Fox River Basin upgradient of Vernon marsh (area is 207 mi^2)			
Recharge	41.89		
Vertical flux across Silurian/Maquoketa boundary	.23	0.94	1.8
Lateral flux across basin boundaries	6.17	7.67	14.7
Surface-water exchange[1]	3.99	40.08	76.7
Pumping		3.68	7.0
			100.2
			[mass balance error = 0.2]

[1]Budget includes exchanges with streams, wetlands, lakes, and quarries but excludes wastewater-treatment plant effluent added to Fox River.

Table 16. Simulated water budgets for nested basins in the coarse-favored model, Upper Fox River Basin.

[Mgal/d, million gallons per day; ΣIN, sum of inflows; mi², square mile; see figure 7 for outlines of basins]

	Inflow (Mgal/d)	Outflow (Mgal/d)	OUT percent of ΣIN
Upper Fox River Basin upgradient of Watertown gage (area is 86 mi²)			
Recharge	16.63		
Vertical flux across Silurian/Maquoketa boundary	.00	0.27	1.4
Lateral flux across basin boundaries	1.43	4.48	22.9
Surface-water exchange[1]	1.51	12.06	61.6
Pumping		2.81	14.4
			100.3 [mass balance error = 0.3]
Upper Fox River Basin upgradient of Waukesha gage (area is 135 mi²)			
Recharge	24.34		
Vertical flux across Silurian/Maquoketa boundary	.10	0.57	1.9
Lateral flux across basin boundaries	3.03	4.80	16.2
Surface-water exchange[1]	2.07	20.94	70.9
Pumping		3.29	11.1
			100.2 [mass balance error = 0.2]
Upper Fox River Basin upgradient of Vernon marsh (area is 207 mi²)			
Recharge	41.89		
Vertical flux across Silurian/Maquoketa boundary	.23	0.94	1.8
Lateral flux across basin boundaries	5.57	6.33	12.0
Surface-water exchange[1]	5.17	42.17	79.8
Pumping		3.42	6.6
			100.1 [mass balance error = 0.1]

[1]Budget includes exchanges with streams, wetlands, lakes, and quarries but excludes wastewater-treatment plant effluent added to Fox River.

6.3 Contributing Areas for Groundwater Sinks

The input and output of the Upper Fox models can be used as input for the particle tracking program MODPATH (Pollock, 1994) in order to delineate groundwater basins based on the simulated flow systems (as opposed to surface-water drainage basins based on topography). For this study, 35,719 particles were released at a 500-ft spacing over the entire model domain at a vertical starting position just below the simulated water table. The effective porosity, an input that affects travel times simulated by MODPATH, was assumed to be 0.2 for the unconsolidated layers and 0.02 for the dolomite layers. Particles were tracked forward under steady-state (2005) conditions until their pathlines terminated at a sink—a surface-water feature, quarry, pumping well, or model boundary. The starting locations of particles discharging to each tributary system within the Upper Fox River Basin were tabulated, as were particles discharging directly to the main trunk of the Fox River. Particles discharging to minor unnamed tributaries connected directly to the Fox River (fig. 6*A*) and to riparian wetlands along the Fox River (fig. 18) were grouped with particles terminating at the Fox River itself. Finally, the starting locations of particles discharging to quarries and the high-capacity pumping wells inserted in the models were separately tabulated.

By mapping the starting locations of the particles for each kind of sink, it is also possible to map areas where groundwater contributes to the sink discharge. The implied groundwater contributing basins to surface-water features as well as the contributing areas to man-made features (wells and quarries) are very similar for fine-favored (fig. 38*A*) and coarse-favored (fig. 38*B*) models. The groundwater contributing basins tend to extend upgradient from the Upper Fox River drainage basin boundary to the west and fall short of the drainage boundary to the east, probably because of tendency of groundwater to flow from west to east, especially in the Silurian. Quarries and wells divert groundwater from its natural surface-water sinks in selected areas.

The calculated median travel time across the groundwater contributing basins associated with surface water is 11.4 years for the fine-favored model and 9.9 years for the coarse-favored model. Simulated travel times are longer to quarries (median time equals 25.4 years for the fine-favored model) and to high-capacity wells (median time equals 63.4 years for the fine-favored model). These results are reasonable given that flow paths are longer to the man-made features. The quarries are excavated at the top of the Silurian dolomite, and most of the well withdrawal is from the dolomite, forcing the discharging groundwater to travel not only laterally but downward.

In addition to the spatial insight derived from the particle tracking results, insight can be gained into the quantity of water flowing to the surface-water network by collecting water-budget terms (table 17). For the Upper Fox River Basin, the total base flow simulated by the fine-favored model is 88.0 ft^3/s and 89.4 ft^3/s for the coarse-favored model. The source of the base flow is partitioned among eight tributary systems (including the headwater portion of the Fox River) as well as minor direct tributaries to the Fox, the main trunk of the Fox, return flow from riparian wetlands, return flow from quarries, and added inflow from WWTPs. The tabulated quantities for both models indicate that about 36 percent of the total base flow to the Fox River is derived from the WWTPs. Keeping in mind that the steady-state simulations correspond to low-flow conditions, one can say that in periods where the supply of water to the Fox River surface-water network is dominated by groundwater flow, about one-third of the flow draining toward the Vernon Marsh is treated wastewater.

Table 17. Simulated base flow within the Upper Fox River Basin.

[ft^3/s, cubic foot per second]

	Fine-favored model (ft^3/s)	Coarse-favored model (ft^3/s)
Tributary base flow at confluence with Fox River		
Fox Headwaters	2.1	2.1
Lannon Creek	1.2	1.2
Sussex Creek	5.6	5.8
Poplar Creek	2.5	2.9
Pewaukee Lake and Pewaukee River	10.6	11.3
Pebble Creek	5.9	6.0
Genesee Creek	7.1	7.0
Pebble Brook	7.6	8.0
Sum of base flow from major tributaries	42.5	44.1
Sum of base flow from minor tributaries	.2	.3
Sum of tributary flow to Fox River	42.7	44.5
Net gain of base flow along main trunk of Fox River	7.6	8.1
Contribution of riparian wetlands	3.7	3.4
Sum of return flow from quarries	2.0	1.6
Added flow from wastewater-treatment plants	31.9	31.9
Fox River base flow above Vernon Marsh	88.0	89.4

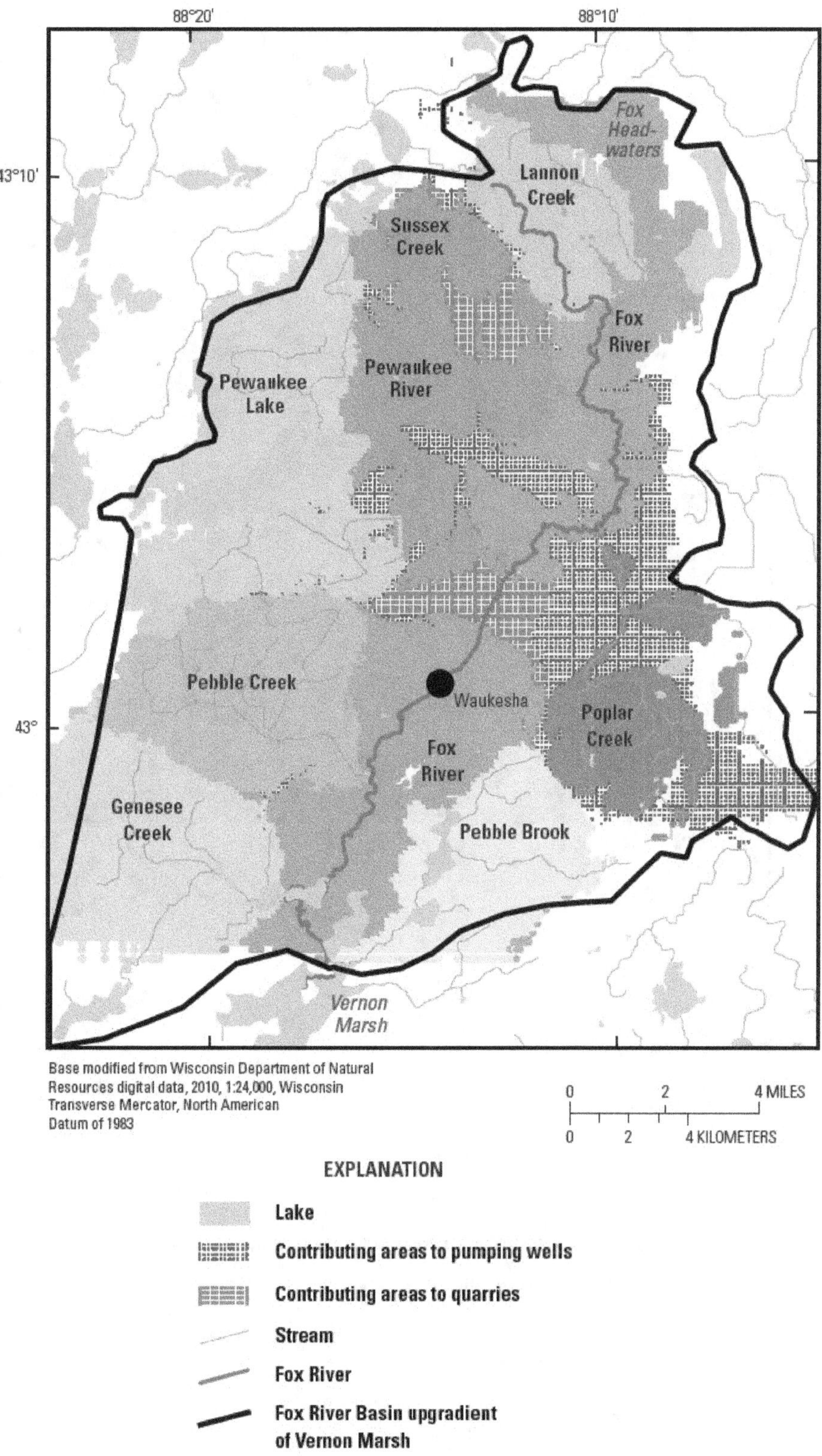

Figure 38*A*. Simulated groundwater basins associated with surface-water features, quarries, and high-capacity wells—fine-favored model.

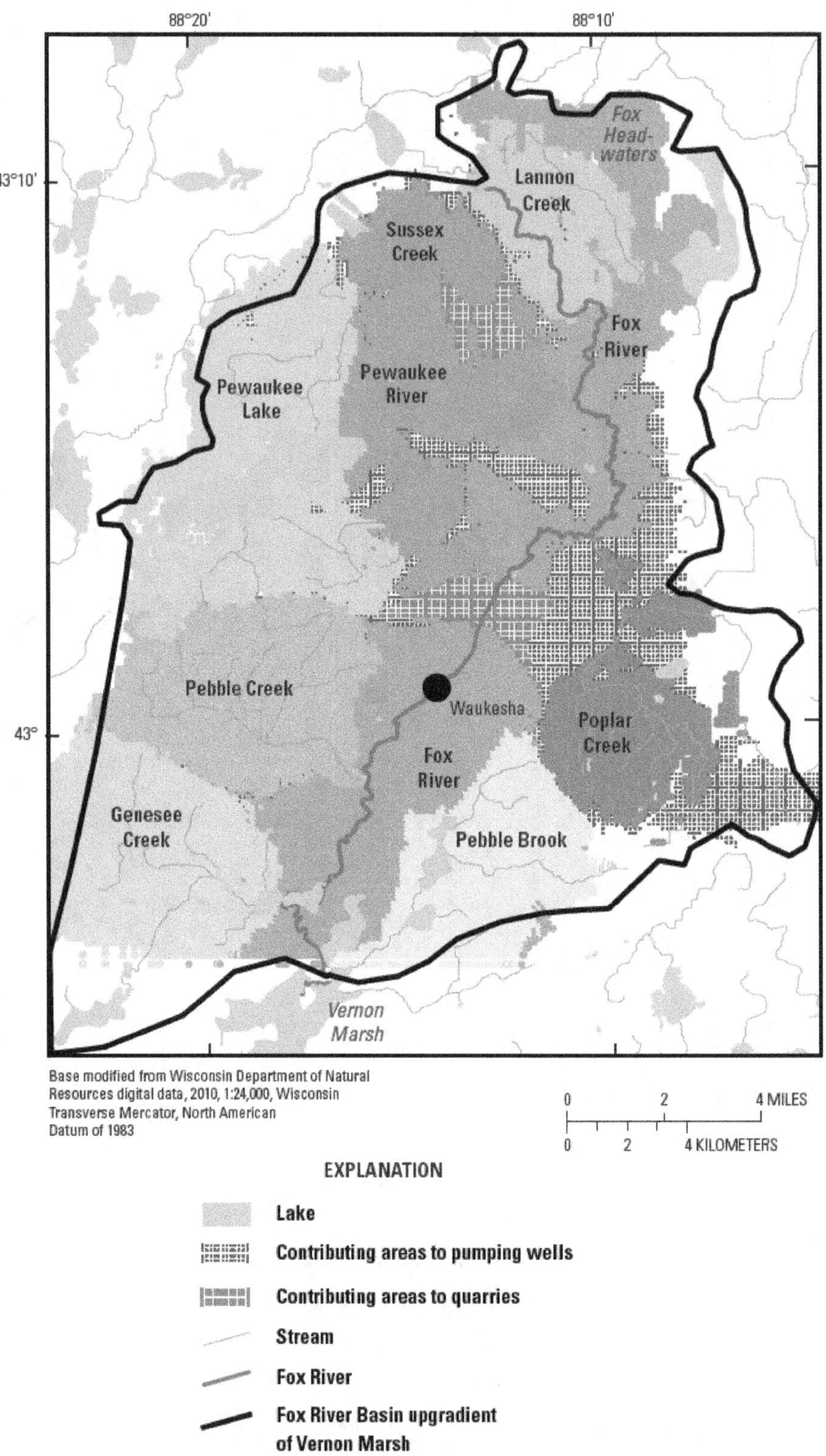

Figure 38*B*. Simulated groundwater basins associated with surface-water features, quarries, and high-capacity wells—coarse-favored model.

One special component of the integrated surface network is Pewaukee Lake. The calculated groundwater contributing basin for the lake extends far beyond its boundaries (green area in fig. 38*A* and 38*B*). The models also simulate the stage of the lake, which partly controls the rate of surface-water outflow to the Pewaukee River. The surface-water outflow is fed by streams flowing into the lake, by groundwater discharge to the lake, and by precipitation on the lake. These inflows are lost to evaporation as well as to surface-water outflow. The tabulated fluxes indicate that all these terms are important in the lake budget (table 18). The surface-water outflow is about 7.52 ft^3/s for the fine-favored model (representing 47 percent of the simulated flow through the lake) and about 7.9 ft^3/s for the coarse-favored model (representing about 48 percent of the simulated flow through the lake). This outflow component is almost one-half of the increase in base flow estimated between the Watertown and Waukesha gages (about 16 ft^3/s). The remainder is derived from base flow to the Pewaukee River and directly to the Fox River.

The lakebed hydraulic conductivity was fixed for all simulations at a value of 0.01 ft/d. Varying its value has very little effect on the simulated lake stage or lake budge. For example, in the case of the fine-favored model, varying its value between 0.001 ft/day to 1 ft/day caused simulated lake stage to change from its calibrated stage of 852.820 ft by no more than 0.004 ft and simulated surface-water outflow to change from its calibrated rate of 7.52 ft^3/s by no more than 0.13 ft^3/s (table 19).

Table 18. Simulated stages and water budgets for Pewaukee Lake.

Simulated results	Fine-favored model	Coarse-favored model
Stage (feet)	852.820	852.833
Inflow (cubic feet per second)		
Precipitation	9.33	9.33
Groundwater inflow	2.53	2.60
Surface-water inflow	4.28	4.54
TOTAL	16.13	16.47
Outflow (cubic feet per second)		
Evaporation	8.45	8.45
Groundwater outflow	.16	.08
Surface-water outflow	7.52	7.94
TOTAL	16.13	16.47

Table 19. Sensitivity of simulated stage and water budget for Pewaukee Lake to lakebed hydraulic conductivity: fine-favored model.

[Kbed, hydraulic conductivity of lakebed; ft/d, foot per day; ft, foot; ft^3/s, cubic foot per second; results for calibrated model repeated from table 18 to facilitate comparison of sensitivity results]

Simulated results	Calibrated model (Kbed = 0.01 ft/d)	Sensitivity model 1 (Kbed = 0.001 ft/d reduced by 0.1x)	Sensitivity model 2 (Kbed = 0.1 ft/d increased by 10x)	Sensitivity model 3 (Kbed = 1.0 ft/d increased by 100x)
Stage (ft)	852.820	852.819	852.818	852.816
Inflow (ft^3/s)				
Precipitation	9.33	9.33	9.33	9.33
Groundwater inflow	2.53	2.20	2.69	2.75
Surface-water inflow	4.28	4.45	4.20	4.17
TOTAL	16.13	15.98	16.21	16.25
Outflow (ft^3/s)				
Evaporation	8.45	8.45	8.45	8.45
Groundwater outflow	.16	.06	.27	.41
Surface-water outflow	7.52	7.47	7.49	7.39
TOTAL	16.13	15.98	16.21	16.25

7. Model Limitations

Model limitations arise from aspects of the conceptual model as well as simplifications inherent in the construction and calibration of the numerical model. This section describes these limitations and related uncertainties.

7.1 Conceptual Model

A major conceptual issue involves the use of water-well driller logs (along with geologic logs) to describe the hydrogeology. Water-well logs are of mixed quality because of variability in the quality of the data collected by drillers. Some investigators prefer to rely on a small sample of high-quality logs only,such as those prepared by geologists (Jansen, 2009), at the risk of introducing a high degree of homogeneity into the geologic interpretation. The large number of available water-well logs partly offsets the mixed quality (Arihood, 2009), but given their uneven spatial distribution, it is inevitable that in some areas few logs are used to interpolate subsurface properties over relatively large distances. In this application, the distribution of logs is comparatively sparse in some focus areas (for example, in the valley of the Fox River where the density of housing is less than elsewhere) (fig. 7).

In order to partly overcome this limitation, two interpolation schemes for mapping hydrogeologic facies in the unconsolidated deposits have been employed (1) connectivity among fine-grained facies is favored and (2) connectivity among coarse-grained facies is favored (figs. 15 and 16). This approach was adopted to allow for a range in the degree to which preferential flow is expressed in the models. Preferential flow, too often neglected in modeling studies or treated by scaling up K parameters to meet calibration targets, is an inevitable feature of highly heterogeneous material such as the glacial deposits characteristic of the Upper Fox River Basin. The detailed cell-by-cell representation of the unconsolidated hydrogeology in this study (along with the spatially refined model mesh in the nearfield) allows direct simulation of preferential flow paths along laterally and vertically oriented horizons to natural discharge zones such as streams as well as man-made discharge points such as pumping wells. However, it is conceded that the fine-favored nor the coarse-favored model incorporates a true representation of the hydrogeology and, that at any specific location, especially in areas of sparse data, the distribution of preferential flow pathways could be exaggerated or suppressed. The use of the Upper Fox models for trial simulations to characterize the likely average response of the aquifers to hypothetical stresses is probably reliable, but without ways to precisely map the subsurface including zones of preferential flow—through site-specific geologic studies and/or other lines of evidence (for example, a dense array of water-level measurements or chemical tracers)—the results of any application for site-specific or design purposes should be treated with caution.

Several other limitations are associated with the conceptual model:

- The lumping of the log descriptions into three texture classes (all fines, mixed, and all coarse) and the lumping of the layer intervals into five facies (dominantly fine, relatively fine, mixed fine and coarse, relatively coarse, and dominantly coarse) are simplifications that might blur differences in hydraulic conductivity properties. For example, assigning gravelly clay and silty sand to the same mixed texture class could conflate relatively impermeable with permeable sediments;

- Individual fracture zones in the Silurian dolomite underlying the unconsolidated material can exert a major control on local conditions of shallow groundwater flow in southeastern Wisconsin (Jansen, 1995), but there is no attempt to characterize them. The Upper Fox models included a weathered horizon at the top of the dolomite (layer 6), but discrete preferential flow zones associated with fracture traces are absent from the models because they are difficult to map without site-specific investigations.

- Stresses on groundwater from domestic pumping are neglected. Although the number of domestic wells is large compared to those in other water-use categories, households generally use relatively small amounts of groundwater and pump it from shallow aquifers; the drawdown cone around each well is commonly buffered by nearby surface water unless recharge is very small (Bradbury and Rayne, 2009). In addition, most of the pumped water is returned to groundwater through onsite septic systems. In Wisconsin, it is estimated that domestic wells account for 23 percent of total groundwater withdrawals (Lawrence and Ellefson, 1982) and about 25 percent of withdrawals in Waukesha County (Southeastern Wisconsin Regional Planning Commission, 2010), but estimates of return rate have been as high as 80 to 90 percent (Topper, 2007; Cherkauer, 2007). These estimates of withdrawal and return suggest omission of domestic wells from the database underestimates total withdrawals by about 5 percent.

- The base simulations of the Upper Fox River Basin models assume steady-state conditions incorporating recharge estimates and edge boundary fluxes corresponding to average conditions for the basin over the last several decades, pumping rates corresponding to 2005, and estimates of WWTP effluent and lake discharge based on data collected after 2005. It is difficult and perhaps not feasible to construct the models based on fluxes corresponding to a single year. Some error arises from the time mismatch in fluxes, although the fact that all the fluxes correspond to recent years probably means the error is small. A related error is introduced by using a steady-state solution to simulate basin behavior, thereby neglecting the role played by changes in storage. In fact, the modeled groundwater system is always responding to changes in stresses inside and outside the model domain, including not only boundary fluxes and pumping, but also variations in recharge. One way to assess the importance of transient effects is to examine the relative importance of storage release simulated by regional models, which include the Upper Fox River Basin area for the timeframe around 2005. The SEWRPC model was used to simulate transient conditions involving changes in pumping (but not recharge) from 2005 to 2010. For the area corresponding to the Upper Fox River Basin models, simulated storage release to groundwater represented 1.76 percent of total sources, and simulated storage gain from groundwater represented 0.05 percent of total sinks. The Lake Michigan Basin model also was used to simulate transient conditions responding to changes in pumping and recharge for the Upper Fox River domain between 2000 and 2005, the last stress period considered in that regional study (Feinstein and others, 2010). For the area of concern, storage fluxes represented 1.42 percent of sources and 0.13 percent of sinks. These results suggest that the neglect of transient effects is a minor source of error.

- Application of the Upper Fox River Basin models to scenario simulations, which change stresses such as pumping or recharge, could invalidate the specified flux boundary conditions at the domain edges. These conditions correspond to 2005 conditions (as reproduced by the SEWRPC regional model). One check on the severity of the stress on the boundary for a scenario simulation is to consider the drawdown and change in flux at the lateral boundaries of the domain. If the simulated response at the boundaries is very small, then the use of fixed lateral boundary fluxes does not compromise the forecast (for the analysis of an example scenario, see the "Model Application to Hypothetical Well Field" section). Vertical exchange across the bottom of the Upper Fox domain also are fixed in the models, partly reflecting the magnitude of Cambrian-Ordovician aquifer pumping in 2005. Forecasting scenarios that change the balance between deep and shallow pumping could imply a change in the vertical exchange at the boundary between the shallow and deep parts of the flow system. Care should be taken to evaluate the effect of any boundary condition change on the integrity of the simulations. Forecasting simulations could benefit from updating the boundary conditions by adding the new stresses to the regional SEWRPC model and re-extracting the implied boundary fluxes.

7.2 Model Construction

Many of the limitations associated with model construction involve the representation of the surface-water network. Although the fine model mesh enhances the numerical reliability of the simulation, little information is available regarding the streambed properties and how they change from point to point. The use of only three zones to characterize the streambed conductance (as a function of three zones of streambed hydraulic conductivity) is a major simplification and the resulting uncertainty could have implications (for example, the reliability of simulating stream losses and gains over local stretches). Streambed conductance can even be considered a transient property, subject to flooding events, which seasonally change the texture and thickness of the bed material in different stretches. However, model results generally are only sensitive to the streambed conductance when it is low, reflecting the accumulation of fine-grained material. More attention is given to this issue in the "Model Application to Hypothetical Well Field" section of the report.

Stream stage has been calculated only through the SFR package for the headwater reaches accounting for about one-half the total length of channel. Fixing the stages in reaches of stream order two or higher on the basis of interpolation from GIS coverages is not an exact process, but it does have the advantage of stabilizing the model solution. Another uncertain input is the outlet elevation of the spillway for Pewaukee Lake. The chosen elevation reflects the reported average elevation of the weir at the outlet, but it is recognized that the LAK package results are very sensitive to this choice because the spillway elevation controls the simulated stage and outflow. The simulated outflow from the lake (averaging 7.7 ft^3/s for the two models) probably carries an uncertainty, based on the constructed rating curve and uncertainty about the spillway elevation of plus or minus 2 ft^3/s.

The presence of simulated groundwater flooding (fig. 37) is partly an artifact of the absence of riparian discharge zones in the model. The Upper Fox River Basin models could be made more complex by using the MODFLOW-2005 UZF package (Niswonger and others, 2005). This package, which simulates flow between the land surface (or bottom of the root zone) and the water table, can be used to quantify flow conditions near streams and lakes, which results in overland flow of groundwater discharge to surface water (Hunt and others 2007b). It is likely that eventual implementation of the method would yield further insight into groundwater/surface-water interactions along the Upper Fox River and its tributaries.

A second set of issues arises from simplifications inherent in the discretization of the model. For example, the representation of the land surface, which is used as a datum for defining the layer bottoms, is discretized on 500-ft centers so that clusters of 16 cells in the model nearfield, where spacing is 125 ft on a side, share common top and bottom elevations. A related issue stems from the nonhorizontal slope of the layer bottoms. The discretization of the unconsolidated material into layers on the basis of the undulations of the land-surface trend does influence the solution by encouraging the simulated flow to follow the land-surface trend along the surface of least resistance represented by the horizontal hydraulic conductivity. However, the average change in elevation in the model domain is on the order of 200 ft to over 40,000 ft or only about 0.3 degrees (even though very evident in cross sections with vertical exaggeration). Bottom slopes less than 15 degrees are not expected to distort the MODFLOW solution (Henry and others, 1998; Mary Hill, U.S. Geological Survey, written commun., May 25, 2011).

A third type of limitation arises from the time period associated with estimates of model inputs that determine sources and sinks. Whereas the inflow from recharge reflects long-term average conditions, the model pumping rates correspond to 2005, the model edge boundary conditions reflect both long-term recharge and 2005 pumping, the available lake spillway data are for the 2007–9 period and the WWTP effluent was estimated using 2008–9 data. It is clear that the fluxes associated with the model are not tied tightly to 2005 but are better characterized as approximations of recent conditions.

7.3 Model Calibration

The calibration targets all contain error, which contributes to uncertainty in parameter estimation. These sources of error are discussed for each target type in the "Model Calibration" section of the report under "Calibration Targets". It is difficult to quantify the relative uncertainty among target types, but the availability of multiple sets is important for expanding the number of parameters that can be informed by the calibration process.

The distribution of pumping from wells penetrating multiple layers, assigned as a function of layer transmissivity based on initial K_h values, was not updated to reflect the calibrated K_h values. However, sensitivity simulations for the fine-favored and coarse-favored models part way through the calibration process indicate that the models results are almost completely insensitive to the update in the distribution of pumpage between layers. Of more serious concern, the calibrated parameter values do not support the total target withdrawal rates for the high-capacity wells inserted in the models. The loss (4 to 8 percent of withdrawals, amounting to 0.3 to 0.6 percent of simulated total outflows) suggests some calibration error, especially because the models do not include the additional loss because of drawdown induced by well inefficiencies. However, there is no guarantee that the reported discharge rates are fully accurate (it is possible pumping capacity is reflected rather than actual pumping) and, therefore, no effort has been made to reinterpret calibrated values to support the total reported rates.

An important limitation of the calibration process is the use of bounds to enforce the ordering of K values from the finest to the coarsest unconsolidated facies. This technique balances the power of the PEST algorithms to find a best fit against the geological judgment embedded in the underlying conceptual model. However, the presence of some calibrated values at the upper and lower bounds for the fine-favored and coarse-favored models does confirm that the mapped facies in the Upper Fox River Basin only approximate the subsurface reality.

8. Model Application to Hypothetical Well Field

This section describes a demonstration application of the Upper Fox River Basin models. The models serve as a tool for exploring the possibility of augmenting municipal water supply by means of riverbank inducement from hypothetical shallow wells as a way of minimizing drawdown coupled with recirculation of the water to upstream WWTPs as a way of enhancing sustainability.

8.1 Riverbank Inducement

Over two-thirds of the residents of Wisconsin use groundwater as their source of supply; the proportion is 86 percent in Waukesha County (Southeastern Wisconsin Regional Planning Commission, 2010, table 29). As populations grow, the expanding demand for water will cause increased competition among users and may exceed the groundwater system's supply. Riverbank inducement (RBI) is a process used to augment groundwater supplies in many nearby States and in numerous midwestern urban areas.

Wells are commonly installed adjacent to rivers (or other surface-water bodies) in order to draw water from two sources: (1) intercepting groundwater that previously would have flowed into the river, and (2) inducing water to flow from the river into the aquifer, which is induced infiltration, and RBI is accomplished by reversing the natural hydraulic gradient by pumping groundwater levels below river level. The induced water augments the original groundwater supply and proportionally reduces the amount of drawdown caused by the wells in the pumped aquifer. The reduced drawdown mitigates the effect of pumping on (1) reducing base flow to nearby wetlands and other local surface-water features, and (2) interference with nearby household and high-capacity wells.

Flow in the source river also is reduced by the amount of inducement to a well field, and the magnitude of this impact will depend upon the river's total flow and where the treated effluent is returned. If the wastewater return is downstream from the inducing well field, flows in the river will be reduced between the well field and the return point by the sum of the induced and intercepted water. Alternatively, if the well field's water is transferred to another watershed, the river from which it is induced will have flows reduced below the well field by the sum of the induced and intercepted water. On the other hand, there will be little impact on low flows and the volume of induced water is actually recirculated past the well field if the well field discharge is returned to an upstream WWTP. When the sources of water to the pumping wells include riverbank inducement, then the recirculation of the pumped water from the river to the aquifer to the WWTP and back to the river enhances the sustainability of the system.

8.2 Hypothetical Pumping Scenario Along Fox River

For the demonstration application, areas north and south of the city of Waukesha along the Fox River are selected to test how a riverbank-inducement strategy could be implemented. It is important to emphasize that this scenario is not supported by site-specific evaluations of well capacity, but the hydrogeologic conditions reflect the interpolation methods from compiled logs described in previous sections. For this reason, the application only offers a general idea of the response of the system to riparian wells and is not the basis for an actual well-field design. However, the results of the simulations should be useful for judging the prospects for riparian systems generally in glaciated areas in the Upper Midwest with subsurface conditions similar to the Upper Fox River Basin.

The hypothetical riparian scenario consists of 27 pumping wells distributed along the Fox River over the north/south extent of the model nearfield between the confluence of Sussex Creek and Pebble Brook (fig. 39). The design includes two public wells already installed by the city of Waukesha: WK-11 and WK-12. All 27 wells are located in the model at a lateral distance 125 ft from the Fox River in the riparian zone. The vertical interval for withdrawal is determined by the following logic:

- if 60 ft of unconsolidated material is present, pumping is assigned to the lowest unconsolidated layer with at least 10 ft thickness; and
- if 60 ft of unconsolidated material is not present, pumping is distributed among the combined unconsolidated and Silurian dolomite material in layers 2 through 7.

This filter results in 9 wells extracting from layer 3 (50.1- to as much as 100-ft depth), 8 wells extracting from layer 4 (100.1- to as much as 150-ft depth), and 10 wells extracting from layers 2 to 7 (20.1-ft to variable depth). The unconsolidated sequence is thinner north of Pebble Creek than to the south; consequently most of the 15 wells north of Pebble Creek extend to the dolomite in layers 6 and 7, whereas, the 12 wells south of the creek are limited to the unconsolidated layers 3 or 4.

The target rate for each well is 0.667 Mgal/d, amounting to a total of 18 Mgal/d for the system. Based on the distribution of calculated transmissivity using the calibrated K parameters and layer thickness, about two-thirds of the target pumping is from the unconsolidated material and one-third is from the dolomite bedrock. However, it is an important feature of the MODFLOW-NWT formulation that the *simulated* discharge rate from each well is a function of how much the aquifers in the vicinity of the well can support. Consequently, one of the important outputs of the demonstration exercise is the sustainable pumping achieved relative to the 18 Mgal/d target.

The background pumping from high-capacity wells is maintained at 6.67 Mgal/d. The flux boundary conditions of the model also are not changed in the simulations. If withdrawals from riverbank-inducement wells were intended to substitute for pumping from deeper wells, a more rigorous approach would be to regenerate flux conditions from the regional SEWRPC model with the new pumping distributions. However, simulations conducted as part of the 2010 Southeastern Wisconsin Regional Planning Commission study suggest that the effect of shifting 18 Mgal/d or less of deep pumping to shallow pumping would have only a minor effect on the magnitude of the vertical leakage between the shallow and deep parts of the flow system.

There are three WWTPs in the area of interest. In principle, 5 of the wells could recirculate their water to the upstream Sussex plant, 9 to the upstream Brookfield plant, and 13 to the upstream Waukesha plant. This element of the hypothetical design was not explicitly simulated (by increasing the inflow to the SFR nodes adjacent to the WWTPs) because, given that the stages of the Fox River are fixed in the model, it would have little or no influence on the solution. However, the recirculation of water could be an important element of a real design of a RBI system that would serve to sustain base flow in the river, notably during low-flow periods.

The fine-favored and coarse-favored models were run in steady-state mode to determine the long-term response of the groundwater to the riparian system in terms of drawdown, sustainable pumping, and sources of water to riparian wells, including water induced from the Fox River.

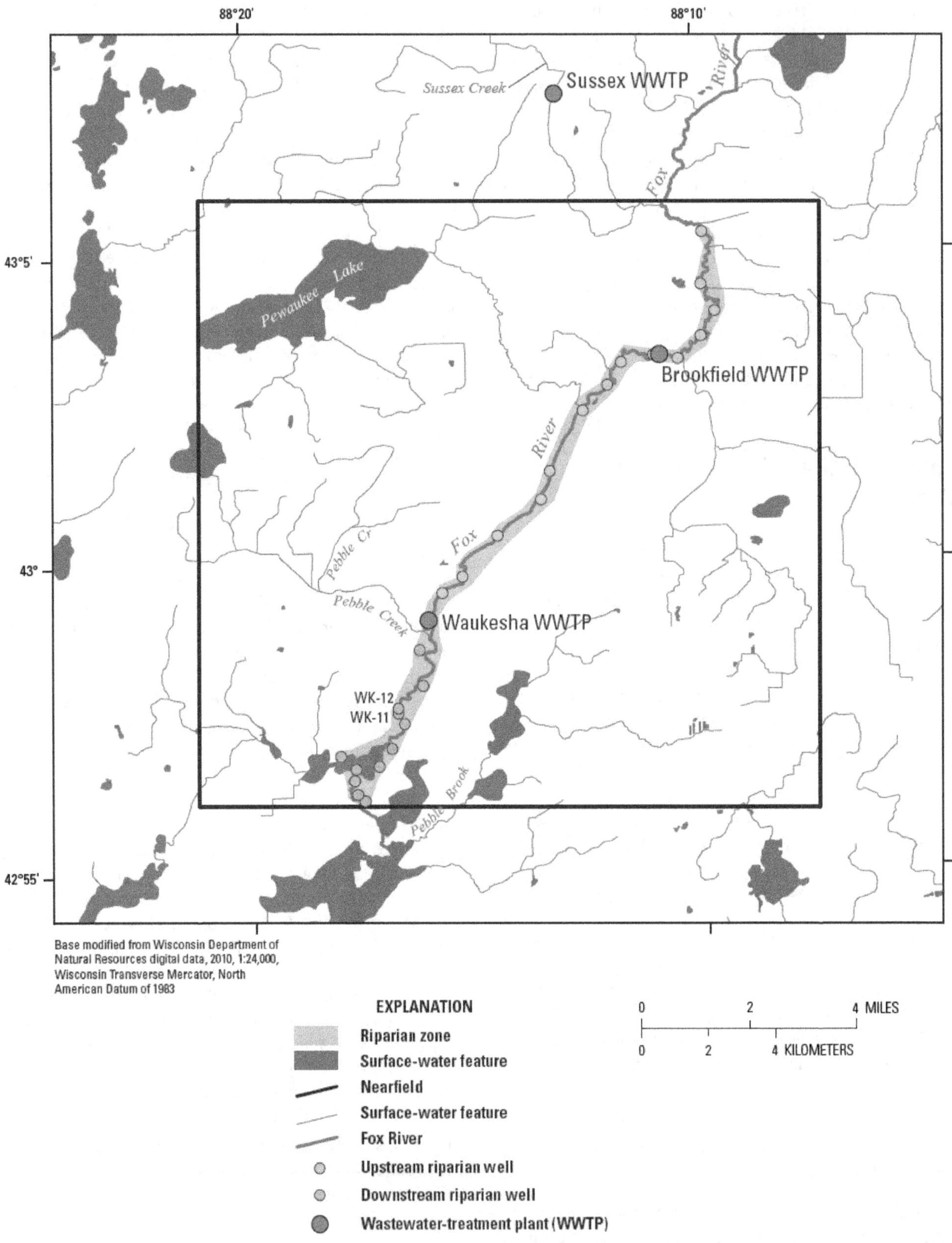

Figure 39. Hypothetical riparian well scenario. (There are 15 riparian wells (orange) located in the area upstream of the confluence between the Fox River and Pebble Creek. There are 12 riparian wells (green) located in the area downstream of the confluence between the Fox River and Pebble Creek The riparian zone (purple) was used for mass balance analysis—see tables 19 and 20.)

The simulated drawdown response at the water table in layer 1 (fig. 40) and at a depth of 50–100 ft in layer 3 (fig. 41) is similar for the two models. Little drawdown occurs outside the Upper Fox River Basin and the drawdown at the edges of the domain is localized and small. Near the Fox River, the maximum drawdown is on the order of 20 ft in layer 1 (fig. 40) and 30 ft in layer 3 (fig. 41). At the water table, drawdown occurs away from the river (because the water level in the stream tends to fix the level of groundwater adjacent to the stream) in areas determined by the relative capacity of the aquifer to transmit water. Note that the simulated drawdown when RBI is ineffective is much larger (see discussion under "Effect of Riverbank Inducement on Drawdown").

The coarse-favored model, as expected, supports more pumping from the riparian system than the fine-favored model. The total pumping sustained from the 27 wells in the fine-favored model is 9.13 Mgal/d (table 20), whereas it is 9.65 Mgal/d in the coarse-favored model (table 21). It is noteworthy that these values are only slightly above one-half the total target rate of 18 Mgal/d. The simulated sources of water to the wells also vary somewhat between the two models. In the fine-favored model, 30.8 percent of the sustained pumping is induced from the Fox; whereas, 34 percent is derived from groundwater that would have discharged as base flow to the Fox in the absence of the riparian system. In the coarse-favored simulation, the water induced from the Fox by the riparian system jumps to 41.3 percent; whereas, 31.7 percent is derived from diverted base flow. This comparatively large difference in the induced component results from the presence of more preferential flow pathways in the coarse-favored model than the fine-favored model, which allow more water in the coarse-favored case to be conveyed from the Fox River laterally and vertically down to the open intervals of the hypothetical wells. The two models taken together, by exploring the uncertainty in our representation of the subsurface, possibly bracket the overall capacity of a real-world riparian system to induce river water to the shallow wells.

Other simulated sources of water for the fine-favored and coarse-favored models include water induced from other streams within the Upper Fox River Basin (5.8 and 2.8 percent in the two models, respectively), base flow diverted from other Upper Fox River Basin streams (15.5 and 14.2 percent), water induced and diverted from wetlands, lakes and quarries in the basin (10.7 and 7.2 percent), and lateral inflow across the boundaries of the Upper Fox River Basin (3.6 and 2.5 percent). The mass balance error (calculated by comparing the sustained pumping to the combined sources of water to the wells) is small (0.2 and 0.3 percent, respectively).

A common element of the two simulations is the 15 hypothetical wells in the northern part of the riverbank-inducement system are less successful in inducing water from the Fox River than the 12 wells in the southern part of the system. For the fine-favored simulation, the induced flux represents only 15 percent of the source water for northern wells, whereas, it constitutes 38.6 percent of the source water for southern wells (table 20). The corresponding induced fluxes simulated by the coarse-favored model are higher (27.7 and 46.5 percent) but also indicate relatively more favorable conditions for RBI in the southern part of the system (table 21). The most important reason for this geographic difference, common to both interpretations, is the presence of more sandy facies in the riparian zone to the south accompanied by higher overall transmissivity.

No effort has been made in the hypothetical exercise presented in this report to optimize pumping by extracting more than 0.667 Mgal/d from RBI wells than could support the increase. Alternative simulations with the two models suggest that some of the wells in the southern area could support higher pumping rates, and at those higher rates, on the order of 50 percent of the source water could be induced from the Fox River.

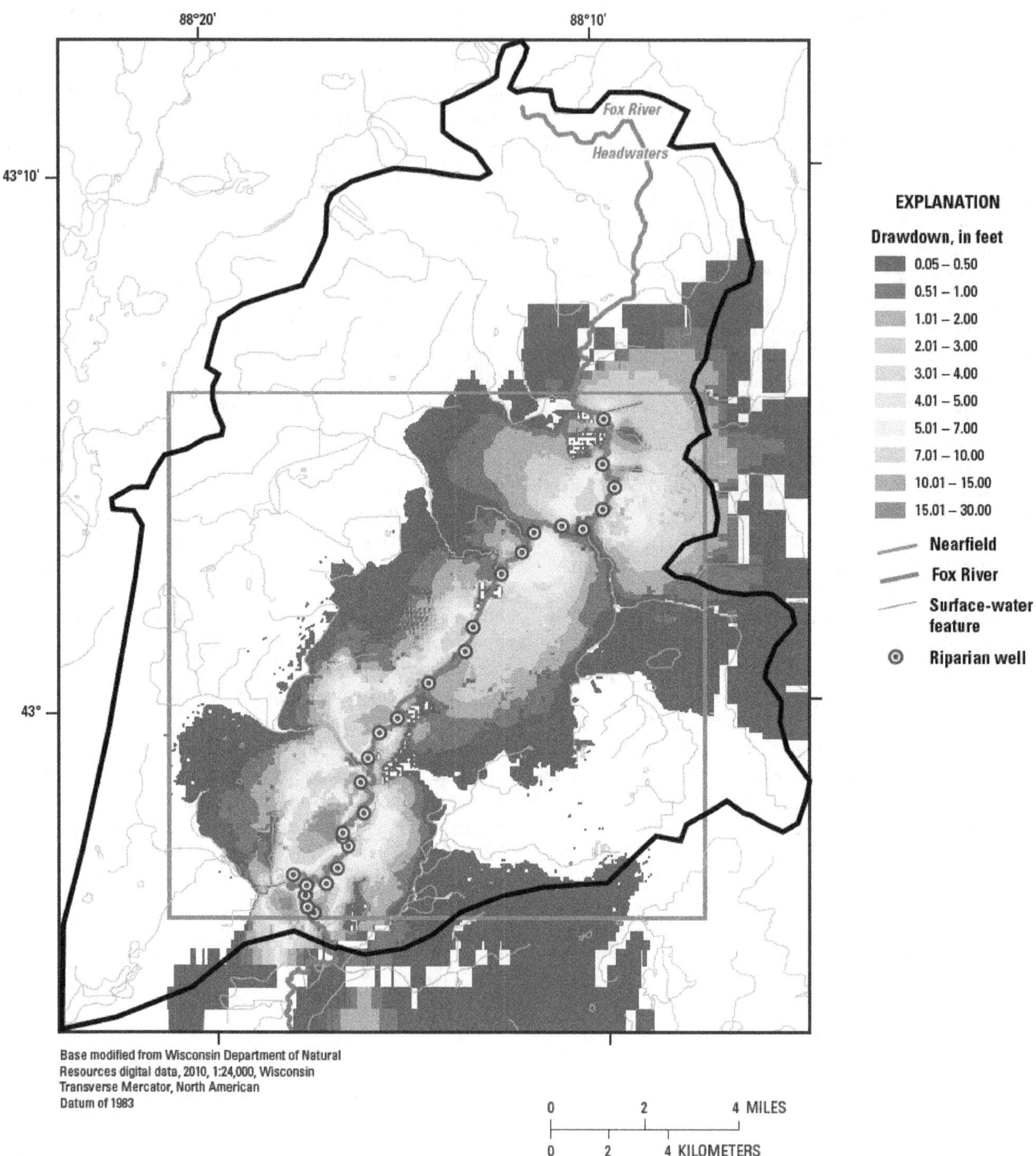

Figure 40*A*. Simulated drawdown from riparian pumping in model layer 1 (water table)—fine-favored model.

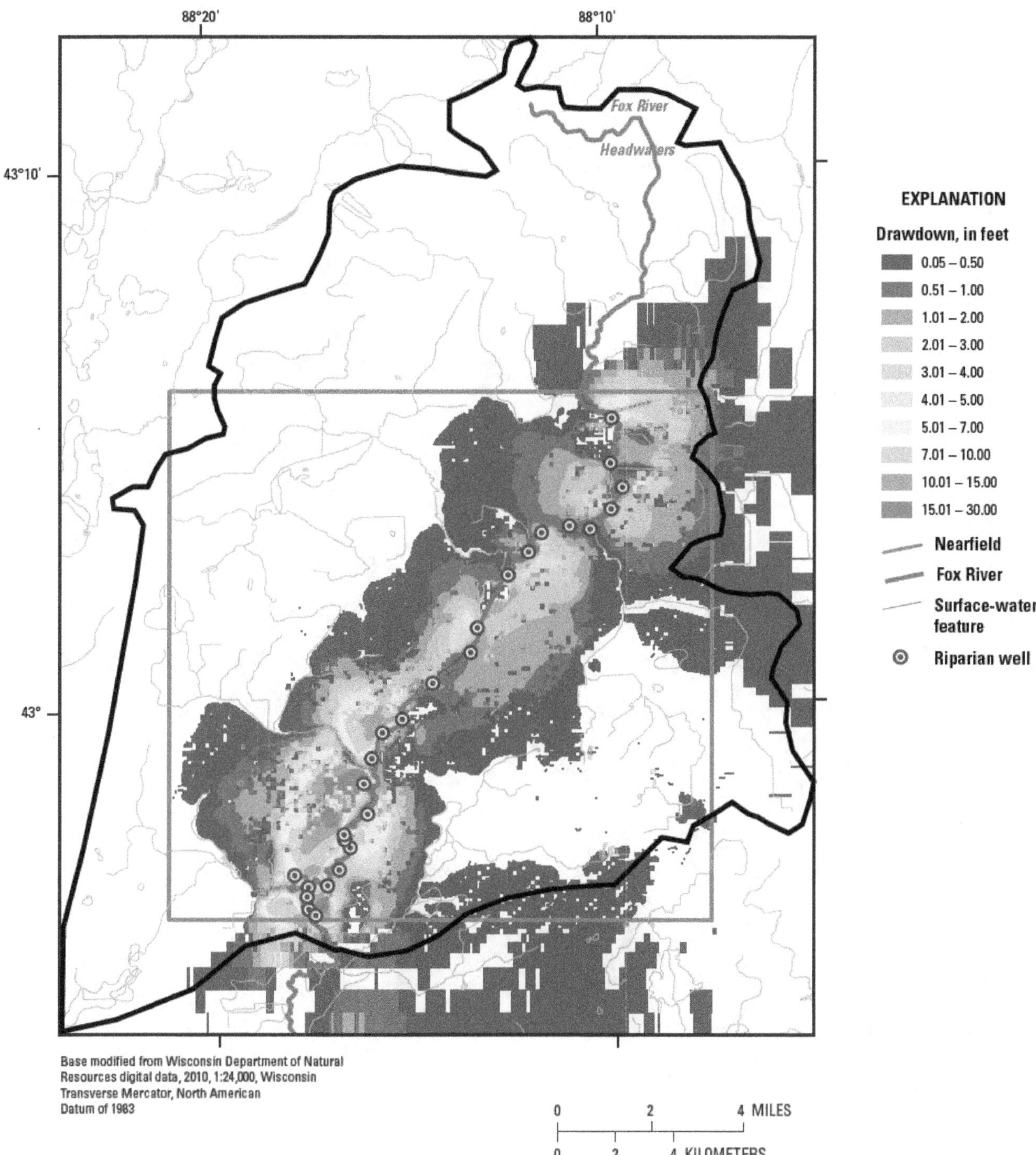

Figure 40*B*. Simulated drawdown from riparian pumping in model layer 1 (water table)—coarse-favored model.

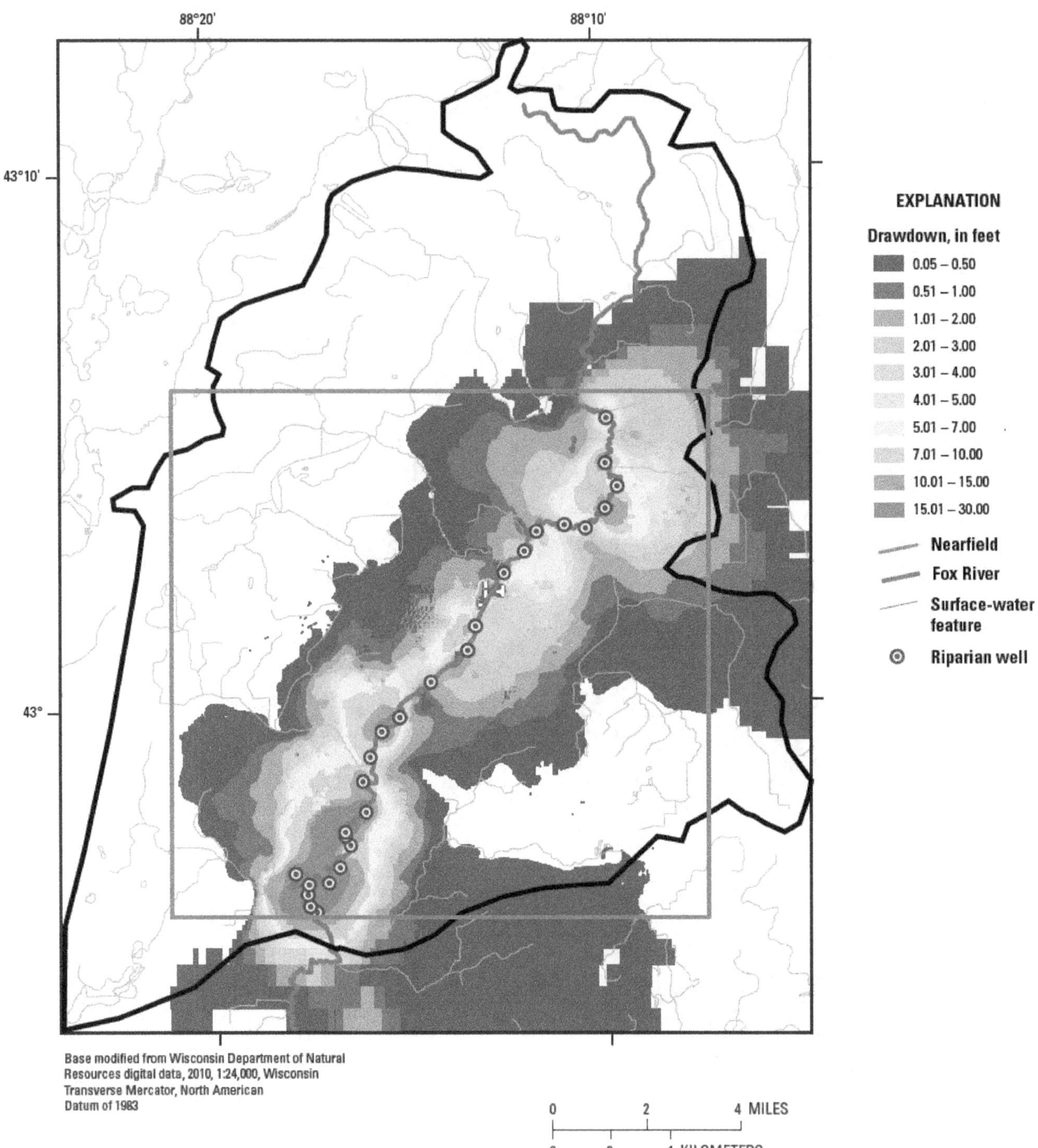

Figure 41*A*. Simulated drawdown from riparian pumping in model layer 3 (50 to100 foot depth)—fine-favored model.

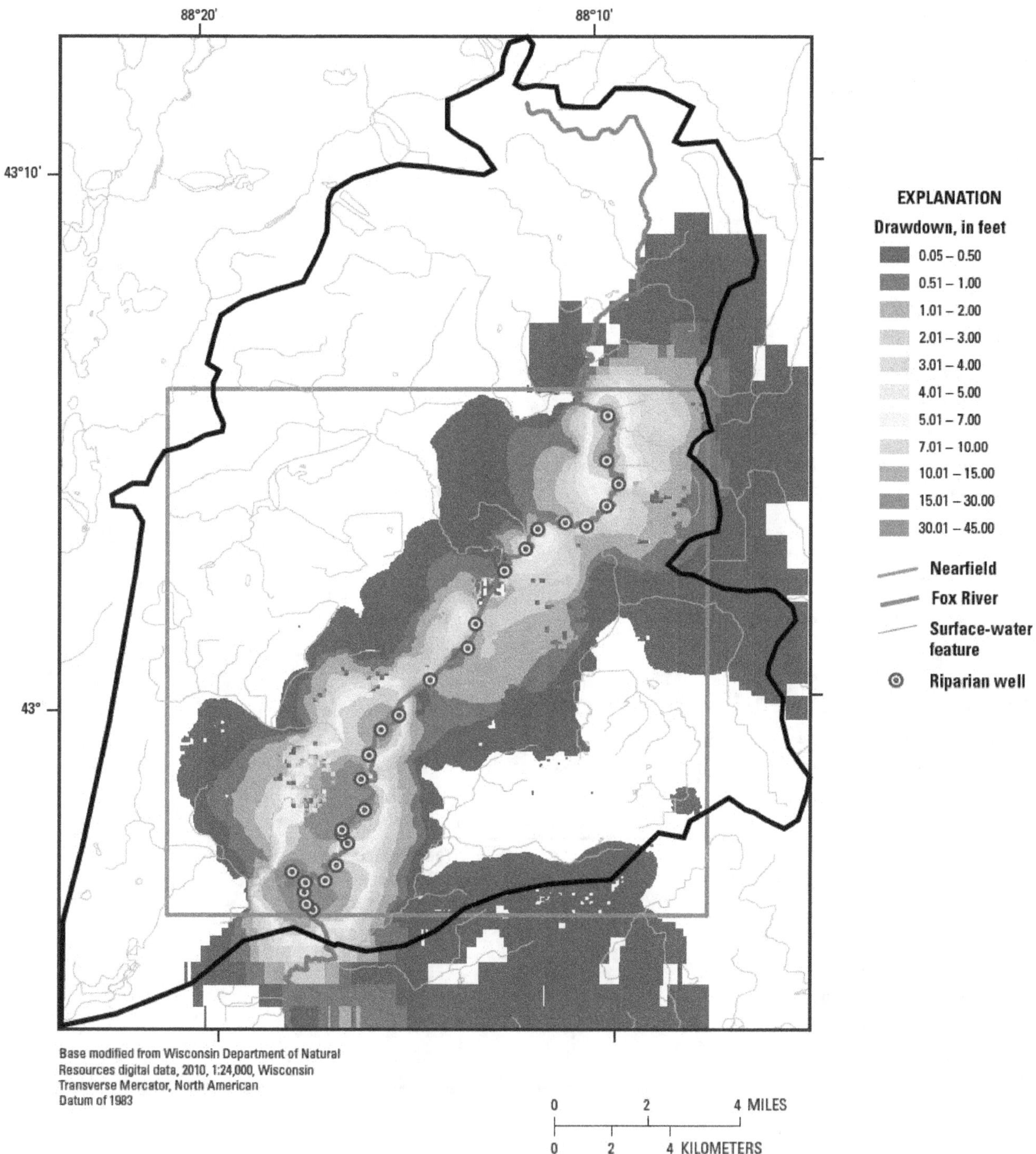

Figure 41*B*. Simulated drawdown from riparian pumping in model layer 3 (50 to100 foot depth)—coarse-favored model.

Table 20. Sources of water to hypothetical riparian wells along the Fox River for fine-favored model.

[Mgal/d, million gallons per day]

Response of all wells			
Number riparian wells simulated	27		
Riparian pumping input (Mgal/d)	18.00		
Riparian pumping sustained (Mgal/d)	9.13		
		Mgal/d	**Sustained pumping (percent)**
Source of water to wells:			
Induced flow from Fox River		2.81	30.8
Diverted flow from Fox River		3.10	34.0
Induced flow from other streams		.53	5.8
Diverted flow from other streams		1.41	15.5
Net induced and diverted flow from wetlands, lakes, quarries		.98	10.7
Lateral flow across Upper Fox River Basin boundary		.33	3.6
TOTAL		9.15	100.2

Response of upstream wells	Upstream area is north of Pebble Creek confluence with Fox River
Number upstream riparian wells	15
Upstream riparian pumping input (Mgal/d)	10.00
Upstream riparian pumping sustained (Mgal/d)	3.04
Induced streamflow to riparian wells (Mgal/d)	.46
Induced from Fox River (percent)	15.0

Response of downstream wells	Downstream area is south of Pebble Creek confluence with Fox River
Number downstream riparian wells	12
Downstream riparian pumping input (Mgal/d)	8.00
Downstream riparian pumping sustained (Mgal/d)	6.09
Induced streamflow to riparian wells (Mgal/d)	2.35
Induced from Fox River (percent)	38.6

Table 21. Sources of water to hypothetical riparian wells along the Fox River for coarse-favored model.

[Mgal/d, million gallons per day]

Response of all wells		
Number riparian wells simulated	27	
Riparian pumping input (Mgal/d)	18.00	
Riparian pumping sustained (Mgal/d)	9.65	

	Mgal/d	Sustained pumping (percent)
Source of water to wells:		
Induced flow from Fox River	3.99	41.3
Diverted flow from Fox River	3.06	31.7
Induced flow from other streams	.27	2.8
Diverted flow from other streams	1.37	14.2
Net induced and diverted flow from wetlands, lakes, quarries	.70	7.2
Lateral flow across Upper Fox River Basin boundary	.24	2.5
TOTAL	9.62	99.7

Response of upstream wells	Upstream area is north of Pebble Creek confluence with Fox River
Number upstream riparian wells	15
Upstream riparian pumping input (Mgal/d)	10.00
Upstream riparian pumping sustained (Mgal/d)	2.68
Induced streamflow to riparian wells (Mgal/d)	.74
Induced from Fox River (percent)	27.7

Response of downstream wells	Downstream area is south of Pebble Creek confluence with Fox River
Number downstream riparian wells	12
Downstream riparian pumping input (Mgal/d)	8.00
Downstream riparian pumping sustained (Mgal/d)	6.98
Induced streamflow to riparian wells (Mgal/d)	3.24
Induced from Fox River (percent)	46.5

8.3 Sensitivity of Results to Streambed Hydraulic Conductivity Values

In this study, the calibration process did not generate reliable estimates for the zonal values of streambed hydraulic conductivity because of insensitivity to the available target types. In general, it is difficult to estimate this parameter (or the related parameter, streambed conductance) from model calibration because most of the head loss in the groundwater system occurs within the aquifer itself; the streambed typically does not offer enough resistance to strongly influence the overall flow system. However, the response to some stresses in the vicinity of surface water (for example, the drawdown response to riparian pumping) are potentially more likely to be affected by the nature of the streambed. In the absence of extensive field data, one way to test hypotheses about its influence is to conduct a sensitivity analysis by varying its properties and recording the effect on drawdown around riparian wells.

The conductance (inverse to the resistance) of the streambed is a function of its area inside the model cell (corresponding in this study to the mapped channel length multiplied by the width as a function of upstream distance), thickness (assumed to be 1 ft everywhere), and hydraulic conductivity (SFR cells grouped into three zones). To simplify the analysis, only the hydraulic conductivity zonal values are varied (so that streambed conductance is a linear function of hydraulic conductivity values), and only the fine-favored model is tested. The base zonal values generated by the calibration process are reduced by 0.001, 0.01, and 0.1 times, and increased by 3 times (table 22). For example, the calibrated value of 9.56 ft/d (characteristic of a fine to medium sand) for the middle streambed K_v zone is changed in the sensitivity simulations to 0.0095, 0.0956, 0.956 and 28.68 ft/d. The effect of the changes on simulated base flow to the Waukesha gage, groundwater flooding, and calibration statistics for nearfield water levels are calculated with all other inputs equal to the base 2005 model, including the high-capacity pumping rate. The effect on the amount of water induced from the Fox River is calculated with the added withdrawals from the 27-well riparian scenario. The simulations with streambed K set at 3 times and 0.1 times the base streambed K_v values show little difference from the results for the base calibrated fine-favored model. The simulation with streambed K_v set to 0.01 times the base value shows sensitivity in terms of a small decrease in base flow, more than doubling of the number of flooded cells, and some deterioration of the calibration statistics; for the scenario simulation, riverbank inducement as a source of water to riparian wells decreases from 31 to 25 percent (table 22). The effects are considerably more severe when the streambed K is set to 0.001 times the base value. For example, rate of riverbank inducement decreases to 6 percent of well sources, less than one-fifth its base scenario value. It is noteworthy that for the most sensitive simulation, the K_v value assigned the middle streambed zone is almost exactly equal to the K_v assigned the fine-dominated facies for the fine-favored model (0.0095 ft/d). The Fox River channel is assigned to the middle K_v zone (fig. 20). Therefore, this sensitivity simulation effectively treats the Fox River bed as fine grained everywhere (perhaps the texture of a clayey silt). If that were the makeup of the bed for the entire channel, then the analysis implies that RBI would be largely ineffective.

Observation and geophysical evidence indicates that at least in the vicinity of water-supply wells WK-11 and WK-12 (fig. 28), the streambed is a mixture of fine-grained and coarse-grained sediment, with sandy and gravelly material dominant in some stretches (Baierlipp and Kean, 2011). This direct evidence suggests it is unrealistic to argue that the For River streambed generally poses strong resistance to groundwater/surface-water exchange. However, it does open the possibility that the exchange is uneven and associated with pathways of preferential flow.

8.4 Effect of Riverbank Inducement on Drawdown

A riverbank-inducement system produces less drawdown than an extraction system isolated from surface water where riverbank inducement does not occur. To quantify the difference, it is convenient to compare the drawdown for the hypothetical 27-well riparian scenario with the base calibrated parameters to the drawdown for the same scenario with the streambed hydraulic conductivity reduced by 0.001 times. Although the latter simulation is out of calibration and although a small amount of water is still induced from the Fox River, it serves as an approximation of drawdown conditions that would be obtained if the system had no access to streamflow as a source of water to the hypothetical wells. Consider the sustained withdrawal and drawdown simulated for the downstream part of the RBI system (fig. 42). The withdrawal is close to 6 gal/min for the set of 12 wells for the two cases, but the drawdown pattern is considerably different between the two cases. In the base scenario simulation, the maximum drawdown is on the order of 25 ft, and the drawdown cone greater than 5 ft is roughly a mile from east to west. In the sensitivity scenario simulation, the maximum drawdown is on the order of 90 ft, and the drawdown cone expands to a width of more than 3 mi. In the base case, riverbank inducement contributes nearly 40 percent of the well water, but it is only about one-fifth of that for the sensitivity case. In the end, this comparison suggests that if site-specific conditions are favorable, riparian pumping is likely to minimize reduction of base flow to surface-water features in the valley and limit drawdown interference between public-supply wells and nearby household wells.

Table 22. Sensitivity of scenario results to streambed hydraulic conductivity.

[Scenario results are compared to calibrated simulation for fine-favored model, shown in boldface; K_v, vertical hydraulic conductivity; ft/d, foot per day; ft^3/s, cubic foot per second, ft, foot; Mgal/d, million gallons per day; ---, not applicable; x, multiplied by]

Run	Fox riverbed K_v (ft/d)	Change to riverbed K_v	Simulated base flow at Waukesha gage (target = 49.6 ft^3/s)	Percent of nearfield with more than 3 ft flooding	Calibration: mean error (ft) Water levels in model nearfield	Calibration: mean absolute error (ft) Water levels in model nearfield	Sustained riparian pumping input = 18 Mgal/d	Well discharge induced from Fox River (percent)
Calibrated-coarse	7.75	---	46.48	6.1	3.04	12.48	9.65	41.3
Calibrated-fine	**9.56**	---	**45.38**	**7.6**	**.89**	**12.05**	**9.13**	**30.8**
SEN00-fine	.0095	reduced by 0.001	41.11	35.3	−15.16	21.23	8.73	6.0
SEN0-fine	.0956	reduced by 0.01x	44.25	17.9	−4.26	13.83	9.04	25.0
SEN1-fine	.956	reduced by 0.1x	45.17	9.8	−.25	12.31	9.11	30.0
SEN2-fine	28.68	increased by 3x	45.41	7.4	1.05	12.03	9.13	30.9

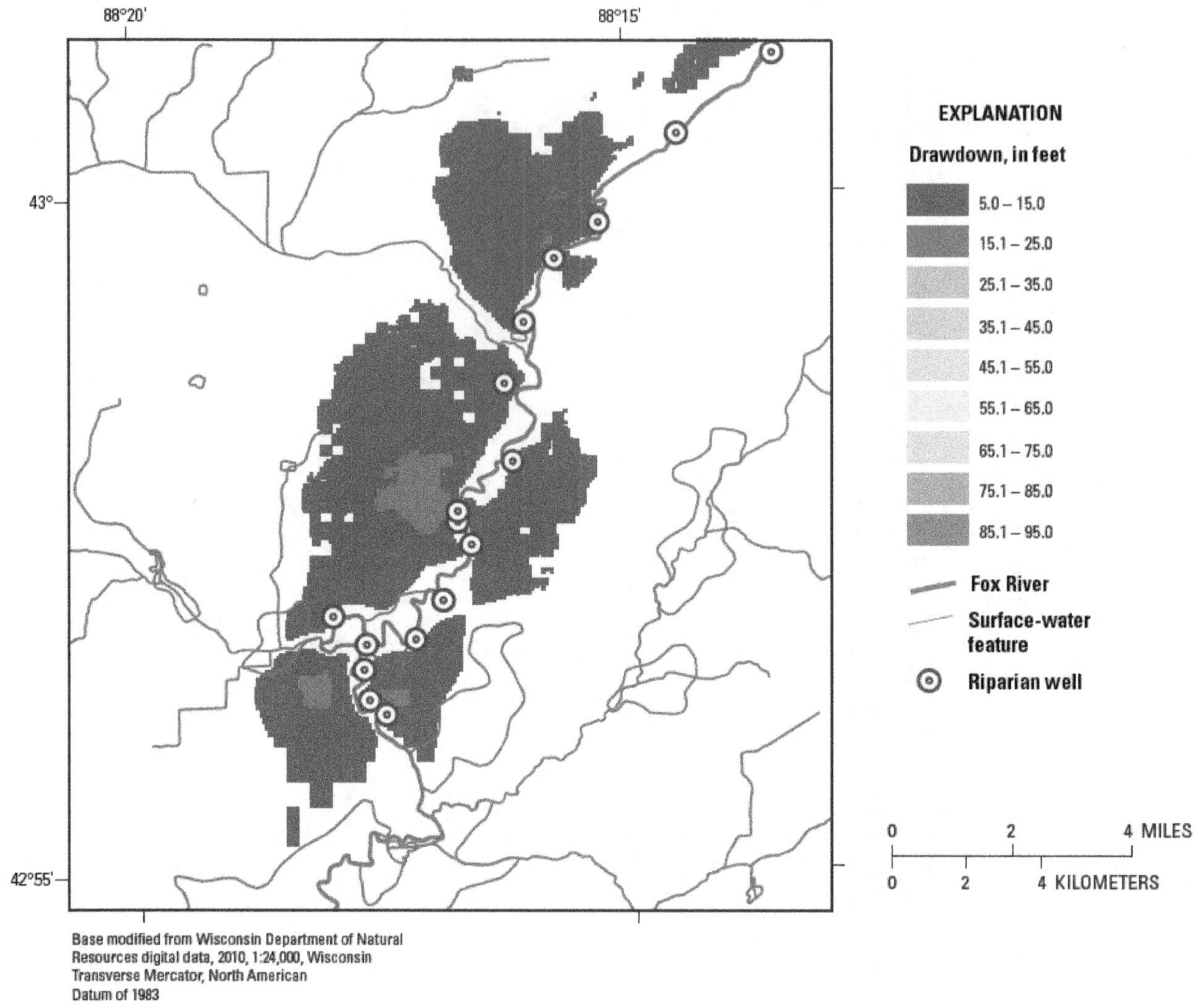

Figure 42A. Effect of degree of riverbank inducement on drawdown in layer 1 (water table) for fine-favored model—riparian simulation allowing riverbank inducement.

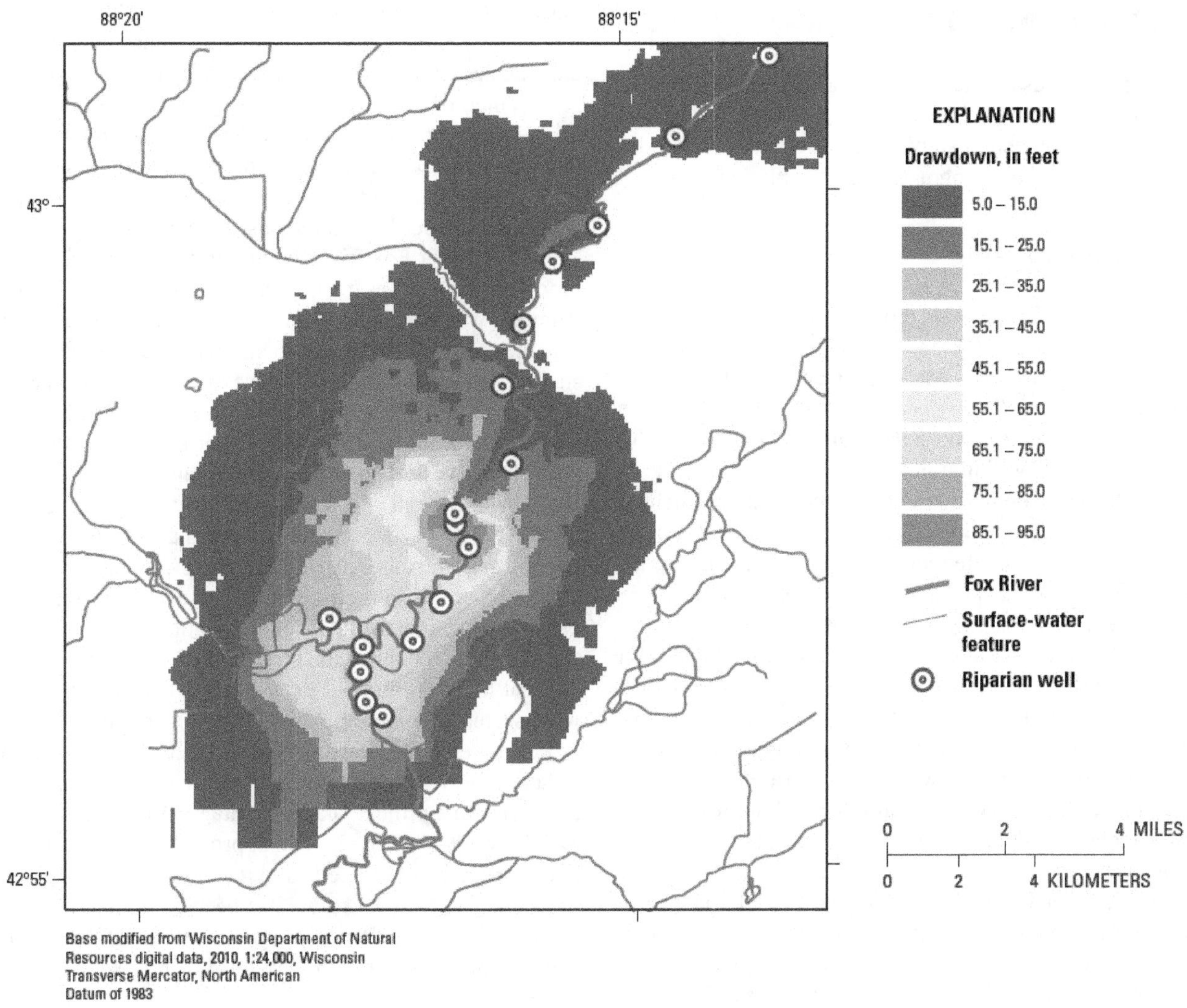

Figure 42*B*. Effect of degree of riverbank inducement on drawdown in layer 1 (water table) for fine-favored model—sensitivity simulation limiting riverbank inducement.

9. Summary and Conclusions

This report discusses the construction, calibration and application of the Upper Fox River Basin groundwater-flow model. It also offers a demonstration of some advanced modeling techniques for simulating unconfined aquifer systems.

9.1 Model Construction

The Upper Fox River Basin model simulates groundwater flow in the area of the Upper Fox topographic basin north of the Vernon Marsh in Waukesha County with the U.S. Geological Survey (USGS) code MODFLOW-NWT. The model domain extends from the source of the Fox River in the southernmost area of Washington County to where the Fox River passes through the northern part of the Vernon Marsh in south-central Waukesha County. The nearfield of the model, characterized by more detailed input and a refined lateral grid spacing 125 feet (ft) on a side, extends north to south between the confluences of Sussex Creek and Pebble Brook with the Fox River and extends west to east from Pewaukee Lake to the tributaries to Poplar Creek near the eastern Waukesha County boundary. All the surface-water features in the nearfield of the model are within the Upper Fox River Basin, including (in addition to the features mentioned above) Pewaukee Creek, Pebble Brook, and Genesee Creek and their tributaries. The model also includes wetland areas and quarries in the Cities of Sussex and Waukesha.

The model, consisting of seven layers, extends vertically from land surface to the top of the Maquoketa shale confining unit which divides the shallow and deep aquifer systems in southeastern Wisconsin. The top five layers represent alluvial and glacial deposits. The bottom two layers represent the Silurian dolomite. The layering is organized by depth. Layer 1 incorporates the top 20 ft of unconsolidated material, layer 2 extends from 20 to 50 ft below land surface, layer 3 from 50 to 100 ft below land surface, layer 4 from 100 to 150 ft below land surface, and layer 5 from 150 ft to the bottom of the unconsolidated material. The number of unconsolidated active layers depends on the thickness of the glacial sequence, which varies from 0 ft in isolated areas where the dolomite is at the land surface to more than 400 ft in the Troy Bedrock Valley in the southern part of the model domain. The weathered top of the dolomite is represented by layer 6 (set to 20 ft thick) and the remainder of the dolomite thickness is assigned to layer 7.

The boundary conditions at the edges of the model correspond to groundwater fluxes, which vary by boundary model cell and by layer. The fluxes into or out of the model domain are derived from the output of the Southeastern Wisconsin Regional Planning Commission (SEWRPC) regional flow model for southeastern Wisconsin developed by the USGS and the Wisconsin Geological and Natural History Survey. The model was updated to 2005 pumping conditions as part of SEWRPC's Water Supply Plan; the results of this 2005 simulation are used to set the flux boundary conditions. In particular, the flux condition at the bottom of the Upper Fox River Basin model, mostly downward, represents the vertical exchange between the shallow groundwater-flow system represented by the sequence of glacial and dolomite sediments (simulated by the Upper Fox River Basin models) and the deep groundwater-flow system incorporating the Cambrian-Ordovician aquifer (represented by the lower boundary condition of the Upper Fox River Basin model).

The Upper Fox River Basin model uses advanced USGS modeling techniques to simulate groundwater circulation from the water table to streams, lakes, wetlands, and quarries. The model also is designed to simulate reduced base flow to surface-water features and induced flow from surface-water features to the groundwater in the presence of pumping. The surface water in the Upper Fox River Basin was routed into a single network within the model by means of the specialized LAK (Merritt and Konikow, 2000) and SFR2 packages of MODFLOW-2005. These packages allow the lake level and the stream stage (particularly headwater stages) to be simulated along with groundwater levels. The inputs to these packages includes LIDAR land-elevation data , characterization of stream channel and streambed features, average rates of effluent inflow reported by the Sussex, Brookfield, Waukesha wastewater-treatment plants (WWTPs), and the bathymetry of Pewaukee Lake as well as lake outlet data provided by the village of Pewaukee. In addition to the advanced surface-water packages, this project takes advantage of the recently released MODFLOW-NWT code, which provides numerical stability in the presence of dry cells and generates a smooth water-table solution. The Newton solver has been applied to allow for more accurate simulation of groundwater-surface-water interactions and the short- and long-term responses of the system to pumping.

The modeling approach focuses not only on incorporating the complete surface-water network, but also on reproducing, at the 125-ft by 125-ft scale of the grid, the heterogeneity of the unconsolidated deposits. Generally, these deposits fall into three bands within the model domain – to the east they are dominated by the clay-rich Oak Creek Formation of the Lake Michigan Lobe; to the west by the more sandy Horicon Member of the Green Bay Lobe; and in the center, over most of the model nearfield, by the mixed deposits of the New Berlin Member of the Lake Michigan Lobe. It is particularly challenging to represent the heterogeneity of the New Berlin Member because repeated episodes of deposition and erosion associated with multiple glacial advances and retreats, along with deposition and downcutting of modern streams, has produced a setting in which the sediment texture changes over very short lateral and vertical distances. A dataset of about 7,000 well logs is the basis of the mapping of five unconsolidated zonal textures or "facies" by model layer—(1) dominantly fine, (2) relatively fine, (3) mixed fine and coarse, (4) relatively coarse, and (5) dominantly coarse, where fine deposits are silts and clays and coarse deposits are sands and gravels. The key point is that this modeling approach does not adopt a single interpretation of the unconsolidated facies but attempts to encompass the inevitable uncertainty by developing two facies models—one which favors the continuity of fine-grained deposits and a second which favors the continuity of coarse-grained deposits. Both the fine-favored and coarse-favored models with their distinct zonation of the unconsolidated deposits are carried forward in the calibration and application phases of the model development.

Other important model inputs are recharge and pumping. A single recharge value is associated with each of 27 subbasins within the Upper Fox River Basin and surrounding watersheds based on studies conducted in southeastern Wisconsin. The values range from 1.5 to 9.5 inches/year. The total recharge applied to the domain of the Upper Fox River Basin model is equal to the total recharge applied to the same area in the parent SEWRPC regional model, implying that the flux edge boundary conditions translated from the regional model to the Upper Fox River Basin model are consistent with the applied recharge. The distribution of pumping from unconsolidated and dolomite deposits is equivalent to the input to the revised 2005 version of the SEWRPC regional model; the dataset amounts to 99 wells pumping 6.7 million gallons/day (1.6 Mgal/d from the glacial aquifer, 5.1 Mgal/d from the dolomite aquifer). Domestic pumping, most of which is returned to the shallow aquifer, is neglected in this model. In the application phase, additional wells are inserted to represent withdrawals added after 2005 or to simulate a hypothetical pumping system.

9.2 Model Calibration

Calibration of the fine-favored and coarse-favored versions of the model was performed by means of nonlinear regression using the program PEST. A variety of target types allowed the calibration process to be sensitive to different parameter sets and to enforce controls on vertical as well as lateral flow. More than a thousand targets sampled from contoured **water levels** mapped by the Wisconsin Geological and Natural History Survey provided information on the horizontal and vertical hydraulic conductivity assigned to the five unconsolidated facies as well as on the underlying dolomite; **estimated base flow** derived using a regression method from streamflow statistics at the Watertown and Waukesha gages on the Fox River acted as a check on recharge; eight **vertical head difference targets** at two locations south of the city of Waukesha derived from site investigations for Waukesha wells helped estimate vertical hydraulic conductivity of the unconsolidated facies; transient **drawdown** reported for an aquifer test conducted on Waukesha well 13, located about one-half mile west of the Fox River between the confluence of Pebble Creek and Genesee Creek, yielded information not only on hydraulic conductivity but also on storage parameters; and, finally, the **land surface** was used as a check to minimize unrealistic groundwater "flooding", and thereby, as a way to constrain hydraulic conductivities and recharge. One parameter set that proved relatively insensitive to the calibration targets was streambed conductance, proportional to the assumed vertical hydraulic conductivity of the streambed. The effect of the uncertainty of the streambed conductance can be evaluated; however, in this case it was done by varying the streambed hydraulic conductivity in the application phase of the modeling.

9.3 Model Application

The separate calibration processes for the fine-favored and coarse-favored models produce distinct parameter inputs for hydraulic conductivity zones, storage parameters, and streambed conductance zones. Given their different structures and inputs, the two models also yield somewhat different results with respect to the pattern of groundwater flow and its connection to surface water in the Upper Fox River Basin above the Vernon Marsh. For example, the water budget for the fine-favored model indicates that groundwater outflow from the Upper Fox River Basin is approximately 77 percent to streams, 15 percent to lateral flow across the basin boundaries, 2 percent to downward leakage to the deep part of the flow system, and 7 percent to wells, whereas, the water budget for the coarse-favored model simulates approximately 80 percent to streams, 12 percent to lateral flow, 2 percent to vertical leakage, and 7 percent to wells.

The application is a steady-state long-term simulation of a hypothetical scenario involving 27 riparian wells open to the shallow aquifer systems, which are located along the Fox River between the confluences of Sussex Creek and Pebble Brook with the Fox River and situated within the model nearfield. The objective of this scenario is to test the concept that a system of riparian wells located downgradient from WWTPs could induce enough water from the river to limit drawdown at the water table away from the river (and the effects of drawdown on surface-water bodies and nearby wells) and also allow for recycling of the water back to the river through the WWTP, thereby supporting a more sustainable water-supply system at least from the standpoint of the physical source of water. It must be emphasized that the hypothetical locations of 25 of the riparian wells are not based on any site-specific investigations but are spaced roughly equally apart and are located 125 ft from the river.

The target withdrawal rate for each well is 0.667 Mgal/d, amounting to a total of 18 Mgal/d for the system. However, an important feature of the MODFLOW-NWT solver is that the simulated withdrawal rate from each well is a function of how much the aquifer in the neighborhood of the well can support. In the event, the models support only a little more than one-half the target rate, but the coarse-favored model sustains more pumping (9.65 Mgal/d) from the hypothetical system than the fine-favored model (9.13 Mgal/d). The presence of more connected preferential flow zones in the coarse case probably accounts for the difference. The simulated sources of water to the riparian wells also vary between the two models.

In the fine-favored case, 31 percent of the sustained pumping is induced from the Fox River. For the coarse-favored simulation, the water induced from the Fox River increases to 41 percent. Among other sources of water to wells, the most prominent is diverted base flow that in the absence of pumping would have discharged to the river.

Sensitivity runs were performed on the fine-favored model to test the robustness of the results for the hypothetical application as a function of changes to streambed hydraulic conductivity (assuming a uniform bed thickness of 1 ft). The calibrated (but insensitive) value in the Fox River channel is equal to about 10 ft/d and corresponds to the hydraulic conductivity of sand. When the streambed hydraulic conductivity values were reduced everywhere by three orders of magnitude so that the Fox riverbed hydraulic conductivity corresponds to the permeability of the dominantly fine facies, riverbank inducement falls from 31 percent to 6 percent of the total discharge from the hypothetical riparian well system. In this connection, it is useful to refer to geophysical data recently collected from the Fox River adjacent to Waukesha wells 11 and 12 that suggests the streambed in the vicinity of the wells is not uniformly fine- or coarse-grained, but rather a heterogeneous mixture of textures, which vary over short distances and allow for the presence of preferential pathways favorable to riverbank inducement.

The modeling techniques employed in this study of the Upper Fox River Basin groundwater-flow system can be applied elsewhere in areas where unconfined glacial aquifers are in connection with surface water. The use of a refined mesh and advanced surface-water MODFLOW packages in conjunction with the MODFLOW-NWT formulation offers a powerful approach for capturing key quantitative features of groundwater/surface-water interactions.

9.4 Concluding Discussion

A key feature of this study is the construction and calibration of two models for the same domain based on different conceptual models of the unconsolidated hydrogeology. The intent is to generate a range of outputs that increase understanding of the uncertainty inherent in model results and predictions. Several lessons can be drawn from the application of this approach to the Upper Fox River Basin.

First, the application of distinct interpolation schemes—one favoring the connectivity of fine-grained sediments and one favoring the connectivity of coarse-grained sediments—yields markedly different realizations of the hydraulic conductivity fields in the unconsolidated layers.

Second, the calibration process tends to blunt the difference between the model results despite the contrasting inputs. Because the results from both models are necessarily compared to the same calibration targets, the parameter estimation algorithm tends to match the targets by balancing the greater volume of coarse material in the coarse-favored model with relatively low hydraulic conductivity values, whereas, the greater volume of fine material in the fine-favored model is offset by relatively high hydraulic conductivity values. As a consequence of this "feedback mechanism," many integrating measures of output are similar for the two models, including water budgets, rates of groundwater/surface-water interactions, areas of simulated upward gradients and upward flows, and simulated areas of groundwater basins associated with tributaries to the Fox River.

Third, probably the greatest difference in the simulated output for the two models occurs in analyzing the response to local stresses, particularly, in calculating the source of water for pumping wells. Directions and rates of flow to wells can be especially sensitive to the presence or absence of preferential flow paths linking a source area to a point of discharge, and probably is the reason the responses of the two models differ somewhat when applied to a hypothetical system of riparian wells aligned along the Fox River. The fine-favored model simulates a more modest connection between the wells and the river, inducing 31 percent of the sustained discharge from the river, than does the coarse model, which induces 41 percent of its sustained discharge from the river.

It is undeniable that carrying two groundwater models forward in this instance to better characterize the range of results and their uncertainty constitutes a large burden in terms of cost and time. Many of the quantitative simulated results are similar for the two models and many aspects of the overall basin system could have been achieved with a single model realization, possibly a compromise between the fine-favored and coarse-favored realizations. However, it is also clear that some applications, such as forecasting local possibilities for water supply, benefit from the use of more than one scheme to eliminate possible bias. Neither model is a true representation of the subsurface, but there is value in taking both sets of results into consideration.

10. References Cited

Aquifer Science & Technology, 2004, Engler site test boring results and recommendations: Letter to Mr. Jeff Detro, City of Waukesha Water Utility, September 10, 2004, 5 p. and attachments.

Aquifer Science & Technology, 2006, Geophysical investigation and test boring activities, Lathers site, Waukesha County, Wisconsin: Letter to Mr. Jeff Detro, City of Waukesha Water Utility, May 25, 2006, 14 p. and attachments.

Aquifer Science & Technology, 2008, Well 13 constant rate pumping test analysis: Letter to Mr. Jeff Detro, City of Waukesha Water Utility, August 11, 2008, 6 p. and attachments.

Aquifer Science & Technology, 2010, Well site investigation report at proposed Lathers site: Submitted to City of Waukesha Water Utility, April 2011, 5 p. and attachments.

Arihood, L.D., 2009, Processing, analysis, and general evaluation of well-driller logs for estimating hydrogeologic parameters of the glacial sediments in a ground-water flow model of the Lake Michigan Basin: U.S. Geological Survey Scientific Investigations Report 2008–5184, 26 p.

Baierlipp, M.S., and Kean, W.F., 2011, [Abstract] A hydrogeological study of the Fox River south of Waukesha, WI: American Water Resources Association, Wisconsin Section, 2011 meeting, Program and Abstracts, p. 40.

Bakker, Mark, and Strack, O.D.L., 2003, Analytic elements for multi-aquifer flow: Journal of Hydrology, no. 271, p. 119–129.

Barnes, H.H., 1967, Roughness characteristics of natural channels: U.S. Geological Survey Water-Supply Paper 1849, 213 p.

Batten, W.G., and Conlon, T.D., 1993, Hydrogeology of glacial deposits in a preglacial bedrock valley, Waukesha County, Wisconsin: U.S. Geological Survey Water-Resources Investigations Report 92–4077, 15 p.

Bradbury, K.R., and Rayne, T.W., 2009, Shallow groundwater sustainability analysis demonstration for the SEWRPC region: A report to the Southeastern Wisconsin Regional Planning Commission, Technical Report Number 48, 39 p.

Buchwald, C.A., Luukkonen, C.L., and Rachol, C.M., 2010, Estimation of groundwater use for a groundwater-flow model of the Lake Michigan Basin and adjacent areas, 1864–2005: U.S. Geological Survey Scientific Investigations Report 2010–5068, 120 p.

Burch, S.L., 1991, The new Chicago model—A reassessment of the impacts of Lake Michigan allocations on the Cambrian-Ordovician aquifer system in Northeastern Illinois: Champaign, Ill., Illinois State Water Survey Research Report 119, 52 p.

Carlson, D.A., 2001, Dependence of conductivities and anisotropies on geologic properties within the near-surface aquifer in Milwaukee, Wisconsin: Unpublished Ph.D. dissertation (Geosciences), University of Wisconsin-Milwaukee, 768 p.

CH2MHill in association with Rukert and Mielke, 2002, Report on future water supply: prepared for Waukesha Water Utility, Executive Summary and eight sections.

Cherkauer, D.S., 2004, Quantifying ground water recharge at multiple scales using PRMS and GIS: Ground Water, v. 42, no. 1, p. 97–110.

Cherkauer, D.S., 2007, Simulating the role of domestic wells in the ground-water system of southeastern Wisconsin [abs.]: 31st Annual Meeting of the American Water Resources Association—Wisconsin section, The future of Wisconsin's Water Resources—Science and Policy, Wisconsin Dells, Wis., March 1–2, 2007, p. 51.

Cherkauer, D.S., and Ansari, S.A., 2005, Estimating ground water recharge from topography, hydrogeology, and land cover: Ground Water, v. 43, no. 1, p. 102–112.

Doherty, John, 2008a, PEST, Model Independent Parameter Estimation—User manual (5th ed.): Brisbane, Australia, Watermark Numerical Computing, accessed October 1, 2009, at *http://www.pesthomepage.org/Downloads.php*.

Doherty, John, 2008b, PEST, Model Independent Parameter Estimation—Addendum to user manual (5th ed.): Brisbane, Australia, Watermark Numerical Computing, accessed October 1, 2009, at *http://www.pesthomepage.org/Downloads.php*.

Doherty, John, and Hunt, R.J., 2010, Approaches to highly parameterized inversion—A guide to groundwater model calibration using PEST: U.S. Geological Survey Scientific Investigations Report 2010–5169, 59 p.

Dunkle, K.M., 2008, Hydrostratigraphic and groundwater flow model—Troy Valley glacial aquifer, southeastern Wisconsin: Thesis (M.S.), University of Wisconsin-Madison, 118 p.

Eaton, T.T., 2002, Fracture heterogeneity and hydrogeology of the Maquoketa aquitard, southeastern Wisconsin: Unpublished Ph.D dissertation (Geology and Geophysics), University of Wisconsin-Madison, 211 p.

Eaton, T.T., Bradbury, K.R., and Evans, T.J.,1999, Characterization of the hydrostratigraphy of the deep sandstone aquifer in southeastern Wisconsin—Final report to the Wisconsin Department of Natural Resources: Wisconsin Geological and Natural History Survey Open-File Report 1999–02, 30 p.

Feinstein, D.T., Dunning, C.P., Juckem, P.F., and Hunt, R.J., 2010, Application of the local grid refinement package to an inset model simulating the interactions of lakes, wells, and shallow groundwater, northwestern Waukesha County, Wisconsin: U.S. Geological Survey Scientific Investigations Report 2010–5214, 30 p.

Feinstein, D.T., Eaton, T.T., Hart, D.J., Krohelski, J.T., and Bradbury, K.R., 2005a, Regional aquifer model for southeastern Wisconsin—Report 1, Data collection, conceptual model development, numerical model construction, and model calibration: Waukesha, Wis., Southeastern Wisconsin Regional Planning Commission, Technical Report 41 [pt. 1], 81 p.

Feinstein, D.T., Hart, D. J., Krohelski, J.T., Eaton, T.T., and Bradbury, K.R., 2005b, Regional aquifer model for southeastern Wisconsin—Report 2, Model results and interpretation: Waukesha, Wis., Southeastern Wisconsin Regional Planning Commission, Technical Report 41 [pt. 2], 67 p.

Feinstein, D.T., Hunt, R.J., and Reeves, H.W., 2010, Regional groundwater-flow model of the Lake Michigan Basin in support of Great Lakes Basin water availability and use studies: U.S. Geological Survey Scientific Investigations Report 2010–5109, 379 p.

Gebert, W.A., Radloff, M.J., Considine, E.J., and Kennedy, J.L., 2007, Use of streamflow data to estimate base flow/ground-water recharge for Wisconsin: Journal of the American Water Resources Association, v. 43, no. 1, p. 220–236, accessed January 24, 2008, at *http://www3.interscience.wiley.com/cgi-bin/fulltext/118544614/PDFSTART.*

GeEx, 1989, Report on the groundwater exploration program in the Troy Bedrock Valley for Waukesha Water Utility, Waukesha Wisconsin: Text, figures and logs submitted December 1989 to Waukesha Water Utility [not paginaed].

Groschen, G.E., Arnold, T.L., Harris, M.A., Dupre, D.H., Fitzpatrick. F.A., Scudder, B.C., Morrow Jr., W.S., Terrio, P.J., Warner, K.L., and Murphy, E.A., 2004, Water quality in the Upper Illinois River Basin, Illinois, Indiana, and Wisconsin, 1999–2001: U.S. Geological Survey Circular 1230, 42 p.

Haitjema, H., Feinstein, D., Hunt, R., and Gusyev, M., 2010, A hybrid finite difference and analytic element model for detailed surface-ground water modeling on a regional scale: Ground Water, v. 48, no. 4, p. 538–548.

Haitjema, H., Kelson, V., and de Lange, W., 2001, Selecting MODFLOW cells sizes for accurate flow fields: Ground Water, v. 39, no. 6, p. 931–938.

Hantush, M.S., and Jacob, C.E., 1954, Plane potential flow of groundwater with linear leakage: Transactions of the American Geophysical Union, v. 35, p. 917–936.

Harbaugh, A.W., 2005, MODFLOW-2005, The U.S. Geological Survey modular ground-water model—The groundwater flow process: U.S. Geological Survey Techniques and Methods 6–A16 [variously paged].

Henry, R.M., Find, O.E., and Guiger, N., 1998, Some grid-related limitations of MODFLOW, *in* Poeter, E., Zheng, C., and Hill, M., eds, MODFLOW'98 conference proceedings: Golden, Colo., USA, October 4–8, 1998, p. 219–226.

Hunt, R.J., Doherty, John, and Tonkin, M.J., 2007a, Are models too simple?—Arguments for increased parameterization: Ground Water, v. 45, no. 3, p. 254–262.

Hunt, R.J., Haitjema, H.M., Krohelski, J.T., and Feinstein, D.T., 2003, Simulating ground water-lake interactions—Approaches, analyses, and insights: Ground Water, v. 41, no. 2, p. 227–237.

Hunt, R.J., Luchette, J., Shreuder, W.A., Rumbaugh, J., Doherty, J., Tonkin, M.J., and Rumbaugh, D., 2010, Using the cloud to replenish parched groundwater modeling efforts: Ground Water, v. 48, no. 3, p. 360–365.

Hunt, R.J., Prudic, D.E., Walker, J.F., and Anderson, M.P., 2007b, Importance of unsaturated zone flow for simulating recharge in a humid climate: Journal of Ground Water, v. 46, no. 4, p. 551–560.

Jansen, J.R., 1995, Ground water exploration by the combined use of fracture trace analysis and azimuthal resistivity in the Silurian dolomite aquifer of eastern Wisconsin: Unpublished Ph.D. dissertation (Geosciences), University of Wisconsin–Milwaukee, 558 p.

Jansen, J.R., and Loughry, J., 2009, Troy bedrock valley aquifer model, Waukesha and Walworth Counties, Wisconsin: Prepared by Ruekert &Mielke for the Southeastern Wisconsin Regional Planning Commission, Memorandum Report Number 188, 48 p.

Jansen, J.R., and Rao, M., 1998, Southeastern Wisconsin sandstone aquifer screening model report: Bonestroo, Rosene, Anderlik and Associates, Inc., 62 p.

Kay, R.T., Arihood, L.D., Arnold, T.L., and Fowler, K.K., 2006, Hydrogeology, water use, and simulated groundwater flow and availability in Campton Township, Kane County, Illinois: U.S. Geological Survey Scientific Investigations Report 2006–5076, 99 p.

Krohelski, J.T., Bradbury, K.R., Hunt, R.J., and Swanson, S.K., 2000, Numerical simulation of groundwater flow in Dane County, Wisconsin: Wisconsin Geological and Natural History Survey Bulletin 98, 31 p.

Lawrence, C.L., and Ellefson, B.R., 1982, Water use in Wisconsin, 1979: U.S. Geological Survey Open-File Report 82–444, p. 11–17.

Linsley Jr., R.K., Kohler, M.A., and Paulhus, J.L.H., 1982, Hydrology for engineers (3d ed.): New York, McGraw-Hill, 508 p.

Mandle, R.J., and Kontis, A.L., 1992, Simulation of regional ground-water flow in the Cambrian-Ordovician aquifer system in the northern Midwest, United States: U.S. Geological Survey Professional Paper 1405–C, 97 p.

Merritt, M.L., and Konikow, L.F., 2000, Documentation of a computer program to simulate lake-aquifer interaction using the MODFLOW ground-water flow model and the MOC3D solute-transport model: U.S. Geological Survey Water-Resources Investigations Report 00–4167, 146 p.

Meyer, S.C., Lin, Y.-F., Roadcap, G.S., and Walker, D.D., 2009, Kane County water resources investigations—Simulation of groundwater flow in Kane County and northeastern Illinois: Illinois State Water Survey Contract Report 2009–07, 425 p.

Muldoon, M.A., Simo, J.A., and Bradbury, K.R., 2001, Correlation of hydraulic conductivity with stratigraphy in a fractured-dolomite aquifer, northeastern Wisconsin, USA: Hydrogeology Journal, v. 9, no. 6, p. 570–583.

Niswonger, R.G., Panday, S., and Ibaraki, M., 2011, MODFLOW-NWT—A Newton formulation for MODFLOW-2005: U.S. Geological Survey Techniques and Methods 6–A37, 44 p.

Niswonger, R.G., and Prudic, D.E., 2006, Documentation of the Streamflow-Routing (SFR2) Package to include unsaturated flow beneath streams—A modification to SFR1: U.S. Geological Survey Techniques and Methods 6–A13, 48 p.

Niswonger, R.G., Prudic, D.E., and Regan, R.S., 2005, Documentation of the Unsaturated-Zone Flow (UZF) Package for modeling unsaturated flow between the land surface and the water table with MODFLOW-2005: U.S. Geological Survey Techniques and Methods 6–A19, 62 p.

Poff, R.J., and Threinen, C.W., 1963, Surface water resources of Waukesha County: Wisconsin Conservation Department, 69 p.

Pollock, D.W., 1994, User's guide for MODPATH/MODPATH-PLOT, Version 3: U.S. Geological Survey Open-File Report 94–464 [variously paged].

Rovey, C.W., 1990, Stratigraphy and sedimentology of Silurian and Devonian carbonates, eastern Wisconsin, with implications for ground-water discharge into Lake Michigan: Unpublished Ph.D. dissertation (Geosciences), University of Wisconsin–Milwaukee, 427 p.

Rumbaugh, J.O., and Rumbaugh, D.B., 2007, Groundwater vistas, version 5: Reinholds, Pa., Environmental Simulations, Inc., 375 p.

Schreüder, W.A., 2009, Parallel PEST using BeoPEST: PEST Conference, November 3, 2009, accessed May 31, 2011, at *http://www.prinmath.com/pest/Schreuder-Nov3.pdf.*

Sibson, R., 1981, A brief description of natural neighbor interpolation *in* Interpolating multivariate data: New York, John Wiley & Sons, p. 21–36.

Southeastern Wisconsin Regional Planning Commission, 2006, A regional land-use plan for southeastern Wisconsin—2035: Planning Report Number 48, 210 p.

Southeastern Wisconsin Regional Planning Commission, 2010, A regional water supply plan for southeastern Wisconsin: Planning Report Number 52, v. 1, 831 p.

Southeastern Wisconsin Regional Planning Commission and Wisconsin Geological and Natural History Survey, 2002, Groundwater resources of southeastern Wisconsin: Southeastern Wisconsin Regional Planning Commission Technical Report 37, 203 p.

Stocks, D.L., 1998, Hydrostratigraphy of the Ordovician Sinnipee Group dolomites, eastern Wisconsin: Unpublished M.S. thesis (Geology and Geophysics), University of Wisconsin–Madison, 192 p.

Streeter, V.L., 1966, Fluid Mechanics: New York, McGraw Hill, 705. p.

Syverson, K.M., Clayton, Lee, Attig, J.W., and Mickelson, D.M., eds, 2011, Lexicon of Pleistocene Stratigraphic Units of Wisconsin: Wisconsin Geological and Natural History Survey Technical Report 1, 180 p.

Tikhonov, A.N., 1963a, Solution of incorrectly formulated problems and the regularization method: Soviet Mathematics—Doklady, v. 4, p. 1035–1038.

Tikhonov, A.N., 1963b, Regularization of incorrectly posed problems: Soviet Mathematics—Doklady, v. 4, p. 1624–1627.

Topper, Ralf, 2007, Memorandum to South Platte River Basin Roundtable—Consumptive/non-consumptive water use by residences using individual sewage disposal systems: Colorado Department of Natural Resources, 3 p.

U.S. Geological Survey, 2010a, National elevation dataset, accessed April 2, 2010, at *http://ned.usgs.gov/.*

U.S. Geological Survey, 2010b, National hydrography dataset, available online at *http://nhd.usgs.gov/*

U.S. Geological Survey, 2010c, NWISWeb database accessed May 2010, at *http://waterdata.usgs.gov/nwis/dv/?referred_module=sw.*

Verruijt, A.,1970, Theory of groundwater flow: London, Macmillan, 200 p.

Waukesha County, 2005, Digital terrain model files: Sheboygan, Wisconsin, Aero-Metric, Inc., on behalf of Waukesha County Department of Parks and Land Use, accessed at *http://maps.waukeshacounty.gov/imf/imf.jsp?site=waukesha.*

Waukesha County Internet Mapping Site, accessed December 2009 at *http://www.waukeshacounty.gov/page.aspx?SetupMetaId=11364&id=12008.*

Wikipedia, 2011, Fox River (Illinois River tributary), accessed April 10, 2011, at *http://en.wikipedia.org/wiki/Fox_River_(Illinois_River_tributary).*

Wisconsin Conservation Department (now the Wisconsin Department of Natural Resources), 1966, Lake survey map of Pewaukee Lake.

Wisconsin Department of Natural Resources, 2009, Water well data—Wisconsin well construction reports plus other related files: Wisconsin Bureau of Drinking Water & Groundwater, created December 6, 2005, CD-ROM.

Wisconsin Geological and Natural History Survey, 2004, wiscLITH—A digital lithologic and stratigraphic database of Wisconsin geology: Wisconsin Geological and Natural History Survey Open-File Report 2003–05, 39 p.

Young, H.L., 1976, Digital-computer model of the sandstone aquifer in southeastern Wisconsin: Southeastern Wisconsin Regional Planning Commission Technical Report 16, 42 p.

Young, H.L., 1992, Summary of ground-water hydrology of the Cambrian-Ordovician aquifer system in the northern Midwest, United States: U.S. Geological Survey Professional Paper 1405–A, 55 p.

Young, H.L., Mackenzie, A.J., and Mandle, R.J., 1989, Simulation of ground-water flow in the Cambrian-Ordovician aquifer system in the Chicago-Milwaukee area of the Northern Midwest, *in* Swain, L.A., and Johnson, A.I., eds., Regional aquifer systems of the United States—Aquifers of the Midwestern Area: American Water Resources Association Monograph Series, no. 13, p. 39–72.

USGS

Feinstein and others—**Development and Application of a GW/SW Flow Model using MODFLOW-NWT in Wisconsin**—Scientific Investigations Report 2012–5108

www.ingramcontent.com/pod-product-compliance
Lightning Source LLC
Chambersburg PA
CBHW080256180526
45167CB00006B/2556

* 9 7 8 1 5 0 0 1 5 4 6 8 4 *